BUSH HERITAGE AUSTRALIA

SARAH MARTIN is the author of *Davis McCaughey: A Life* (2012), *When is Forever? A Quirky Tale of Family Life* (2015), and *A Galapagos Day* (2016), a children's picture book. Sarah's research for this book took her to many remote parts of Australia.

RESTORING NATURE STEP BY STEP

BUSH HERITAGE AUSTRALIA

SARAH MARTIN

BUSH HERITAGE AUSTRALIA

25 YEARS OF CONSERVATION EXCELLENCE

NEWSOUTH

A NewSouth book

Published by NewSouth Publishing
University of New South Wales Press Ltd
University of New South Wales
Sydney NSW 2052
AUSTRALIA
newsouthpublishing.com

First published 2016

10 9 8 7 6 5 4 3 2 1

National Library of Australia Cataloguing-in-Publication entry
Creator: Martin, Sarah, author.

Title: Bush Heritage Australia: Restoring nature step by step / Sarah Martin.

ISBN: 9781742235042 (paperback)
9781742235059 (hardback)
9781742247915 (ePDF)
Notes: Includes bibliographical references and index.
Subjects: Bush Heritage Australia – History.
Natural resources conservation areas – Australia.
Nature conservation – Australia.
Wildlife conservation – Australia.
Land tenure – Australia.
Dewey Number: 333.720994

Internal and cover design: Susanne Geppert

IMAGES: *Front cover* Storm over Boolcoomatta Reserve, SA. Photo: Boris Hlavica. *Back cover* The Great Western Tiers, rising above Drys Bluff, Tas. Photo: Wayne Lawler/Ecopix. *Opposite title page* Rocky tor rises from the plains at Boolcoomatta Reserve, SA. Photo: Wayne Lawler/Ecopix. *Part opening images* Susanne Geppert

Printer: 1010, China

This book is printed on paper using fibre supplied from plantation or sustainably managed forests.

BUSH HERITAGE AUSTRALIA
25 YEARS OF CONSERVATION EXCELLENCE
UNSW AUSTRALIA

Cool, damp glades of tree ferns, Liffey River Reserve, Tas

Photo: Wayne Lawler/Ecopix

Contents

Foreword
A brilliant bush career
by Bob Brown

Charles Darwin's great-great-grandson Chris says that 'giving is what makes you happy'. Perhaps that's why Bush Heritage Australia is a landscape of happiness.

For the ever-growing band of benefactors like Chris, and the volunteers and staff of Bush Heritage, happiness flows from our combined effort to ensure that Australia's unique landscapes, wildlife and ecosystems prosper into the future.

However, as Sarah Martin's account of the first twenty-five Bush Heritage years attests, that effort also involves a lot of team-building, risk-taking and hard work.

In 1990–91 Bush Heritage had 236 hectares of land in two blocks of forest and a debt of $200 000.

Perhaps the biggest favour I did the organisation was to get elected to the Senate in 1996. That got me out of the way of a waiting and willing tranche of innovators. In the following decades they brought in expertise in management, fundraising, ecological (including Aboriginal) wisdom, publicity and savvy political lobbying.

By 2000–01 Bush Heritage's annual income was topping $1 million.

Nowadays Bush Heritage owns more than 1 236 000 hectares of lands and is contracted to help in the protection of another 5 657 000 hectares of Aboriginal lands and seas, while its annual income has reached $20 million.

Impressive as these figures are, they are not the point. For it is in the wildlands of Australia that Bush Heritage's tally really matters. Its profits are in protected grasslands, woodlands, forests, rivers, shorelines and deserts, and their diversity of living creatures.

As our late board member and former National Farmers Federation president Rick Farley commented in 2001, 'What I like most about [Bush Heritage] is its practical approach – if there is important country, go out and buy it. Get around all the red tape regulations by taking control yourself. It is a very simple concept, but devastatingly effective.'

Bush Heritage's tally includes protecting species like the night parrot, once thought extinct, and the ancient red-finned blue-eye – a tiny fish confined to a handful of inland mound-springs – along with rare wildflowers and the ancient stromatolites in the west. These triumphs must be seen against the wider reality of global warming, catastrophic loss of species and wildness, and a human horde now consuming earth's living resources at 150 per cent the rate of their replacement.

As I sit here reading Sarah Martin's history of Bush Heritage, under the snow-capped turrets of Taytitikitheeker (Drys Bluff) and the sliver of a new moon high in the north-west sky, Bush Heritage seems magically successful.

But it wasn't magic. Bush Heritage grew through failures and heartaches as well as successes, all bound up in an optimistic devotion to our living planet.

Consider born-and-bred West Australian Margaret Robertson, whose determined expertise and outreach was vital to Bush Heritage's success story in the west; or Robert Rosen, our first treasurer, who lay awake at night charting the organisation's future; or Greg Blake, a young ecology student and 'uncivilised boy' who emerged from Tasmania to range across the continent as Bush Heritage's first land manager; or Verna Simpson who, among a group of contributing young mothers in the 1990s, set up our national fundraising strategy.

Where might we be had Margaret Atkinson not spotted 8 hectares of cassowary habitat and fan palms for sale in the Daintree rainforest in 1993 to give the still-indebted

Bush Heritage its first land holding on mainland Australia for just $85 000?

Martin charts all the vital contributions, like those of my successor as president, fellow Franklin River campaigner Karen Alexander and, from her, the extraordinary Doug Humann, Chief Executive from 1997 to 2011, and our third president Phillip Toyne ('the perfect president for an era of expansion') who described his job as 'far and away the most satisfying environmental engagement' of his illustrious career.

It is all here in Martin's history, along with tantalising tales like the $6 million breakfast, the one ecologically wonderful property that got away, and the remarkable young woman who rode off on her bicycle after leaving a cheque for more than a million dollars next to Doug Humann's coffee cup.

Yet is anything more inspiring than the volunteers who work on the organisation's properties or the pensioners who send in $20 because they want the future of their great-grandchildren to be greener? Or the fact that in 2016 Bush Heritage has a Science Plan involving fifty-five conservation projects, fifty scientists and fifteen universities?

Looking back I thank that little band of good-hearted friends, not least Judy Henderson, who managed the fledgling Bush Heritage's emergence on the road to such a brilliant career.

Looking forward, I wish Bush Heritage's Chief Executive Gerard O'Neill, its President Louise Sylvan and Board, its staff and its growing band of supporters great success and fulfilment in the exciting years to come.

Congratulations to Sarah Martin for this chronicle of twenty-five brilliant Bush Heritage years.

Now let's raise a glass to the next twenty-five years of Bush Heritage Australia and the happiness of giving!

Ephemeral wetland, Ethabuka Reserve, Qld – a temporary oasis in the desert

Photo: Alistair Dermer

Timeline

1990 Bob Brown buys Liffey River and Drys Bluff blocks (Tas). Australian Bush Heritage Fund formally launched

1991 Australian Bush Heritage Fund incorporated, with Bob Brown as President

1993 Fan Palm (Qld) purchased

1995 Brogo (NSW) purchased

1996 Kojonup (WA) purchased. Bob Brown relinquishes presidency; Karen Alexander becomes President

1997 Friendly Beaches (Tas) and lease of Erith Island (Tas) purchased. Doug Humann becomes CEO

1998 Goonderoo (Qld) and South Esk Pine (Tas) purchased

1999 Tarcutta Hills (NSW) purchased, Burrin Burrin (NSW), Currumbin (Qld) acquired

2000 Karen Alexander retires as President

2001 Carnarvon Station (Qld) purchased. Phillip Toyne becomes President. Agreement reached to relocate Bush Heritage to Melbourne

2002 Chereninup Creek (Gondwana Link, WA) purchased

2003 White Wells Station (WA) purchased and becomes Charles Darwin Reserve

2004 Ethabuka (Qld) purchased; Reedy Creek (Qld) acquired

2005 Relocation to Melbourne completed. Judith Eardley Reserve (Vic) established; Eurardy (WA), Cravens Peak (Qld) purchased

2006 Nardoo Hills 2 (Vic), Yarrabee Wesfarmers (Gondwana Link, WA), Boolcoomatta (SA), Scottsdale (NSW) purchased

2007 Peniup Creek, Monjebup (Gondwana Link, WA), Nardoo Hills 3 (Vic), Yourka (Qld) purchased; Nameless Sylvan (NSW) acquired

2008 Bon Bon (SA) and Edgbaston (Qld) purchased. Bush Heritage fails to purchase Wollogorang–Wentworth (Qld) properties

2009 Monjebup North (Gondwana Link, WA) purchased. Phillip Toyne retires as President

2010 Louise Sylvan becomes President

2011 Oura Oura (Tas) donated by Bob Brown and Paul Thomas. Doug Humann retires as CEO; Gerard O'Neill becomes Chief Executive

2012 Naree (NSW) purchased

2014 Monjebup Creek (Gondwana Link, WA) purchased

2015 Hamelin Station (WA) purchased

2016 Habitat of the night parrot (Qld) purchased, Yantabulla Station (NSW) management agreement under negotiation. Bush Heritage ten-year Science Plan launched at Scottsdale Reserve (NSW)

A spark of life in the red sand, Cravens Peak Reserve, Qld

Source: Bush Heritage Australia

Bush Heritage properties and partnerships, 2016

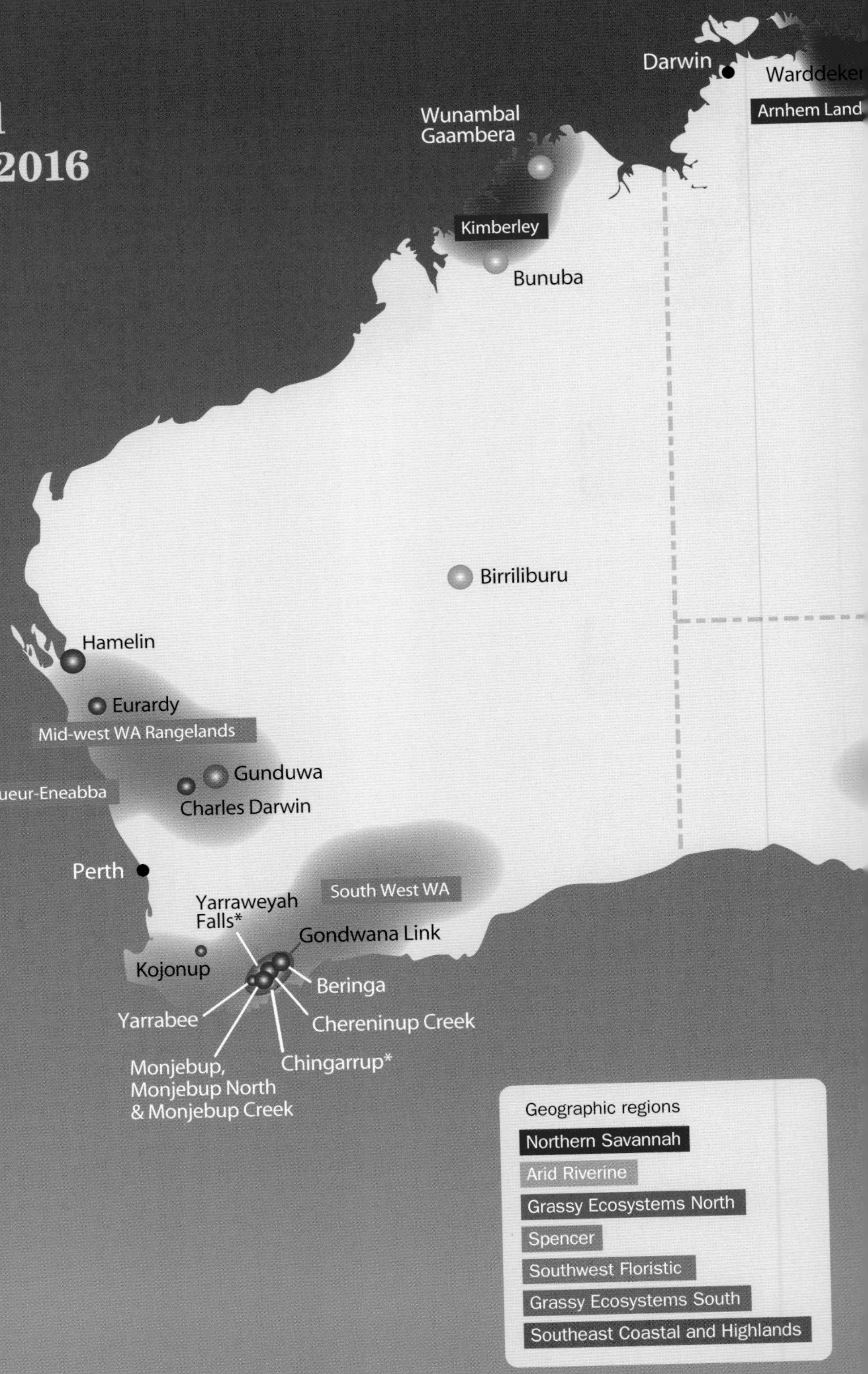

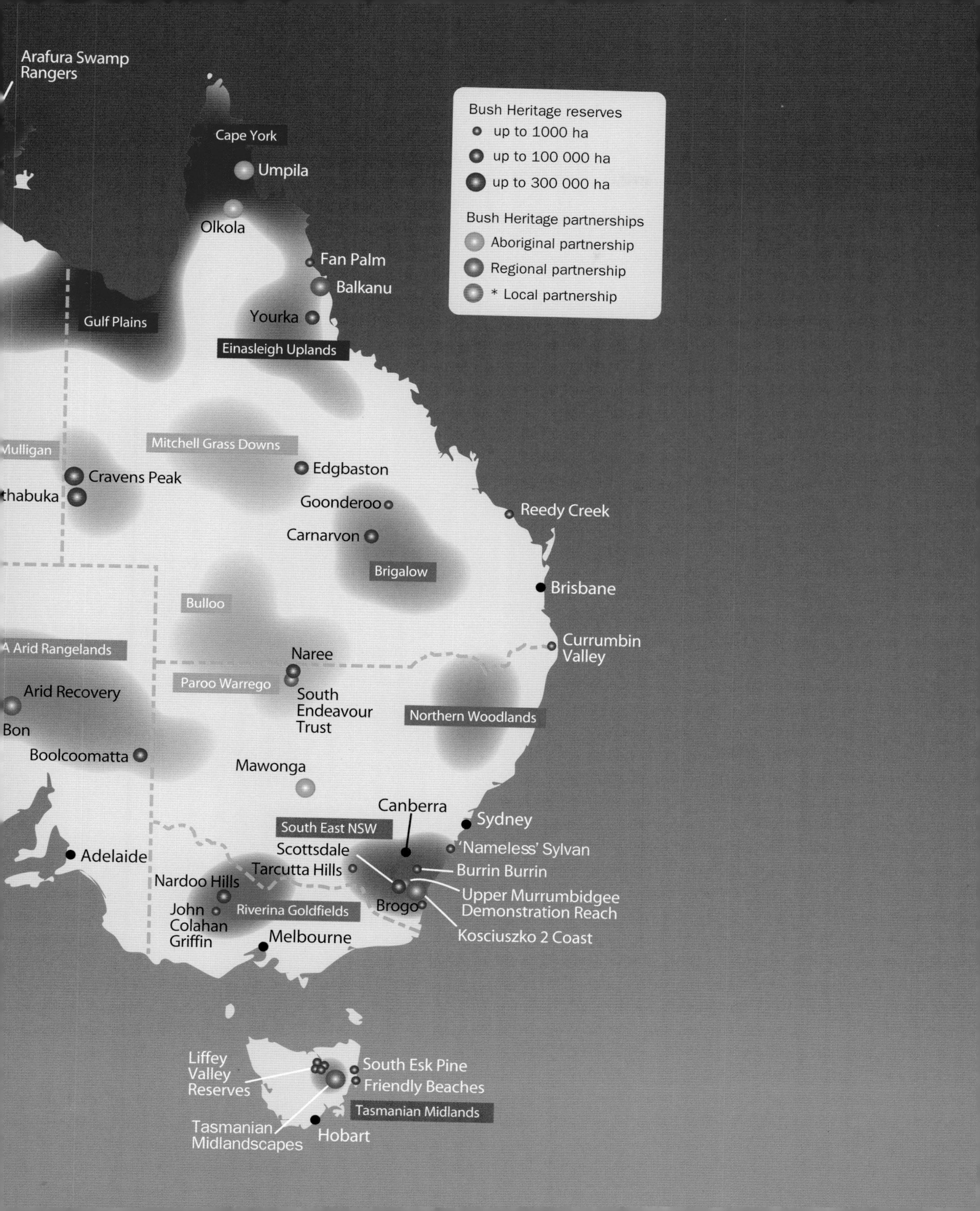

Arafura Swamp Rangers
Cape York
Umpila
Olkola
Fan Palm
Balkanu
Yourka
Gulf Plains
Einasleigh Uplands
Mulligan
Mitchell Grass Downs
Cravens Peak
thabuka
Edgbaston
Goonderoo
Carnarvon
Reedy Creek
Brigalow
Brisbane
Bulloo
A Arid Rangelands
Naree
Currumbin Valley
Paroo Warrego
Arid Recovery
South Endeavour Trust
Northern Woodlands
Bon
Boolcoomatta
Mawonga
Canberra
Sydney
South East NSW
Adelaide
Scottsdale
'Nameless' Sylvan
Tarcutta Hills
Burrin Burrin
Nardoo Hills
Upper Murrumbidgee Demonstration Reach
John Colahan Griffin
Riverina Goldfields
Brogo
Melbourne
Kosciuszko 2 Coast
Liffey Valley Reserves
South Esk Pine
Friendly Beaches
Tasmanian Midlands
Tasmanian Midlandscapes
Hobart
Bush Heritage reserves
up to 1000 ha
up to 100 000 ha
up to 300 000 ha
Bush Heritage partnerships
Aboriginal partnership
Regional partnership
* Local partnership

Prologue

The depression in the sandy soil is slight, the water about 5 centimetres deep. Only the low shadecloth fence encircling it tells me we have at last reached our goal, as I follow Adam's example and step over the barrier, kneel down, and stare into the clear water. Thin strands of grass, brilliant green, wave languidly. Fish glide between them, the biggest less than 4 centimetres long. Glints of red on some and flashes of blue as they turn in synchronised curves highlight their transparency. They are the red-finned blue-eye, the world's most endangered fish. Cut off from the evolutionary loop by a geomorphic accident, they exist today, as they have for millions of years, in a cluster of small desert springs on Edgbaston Reserve in Central Queensland. When discovered in 1990, they were heading for extinction. Today, their chances of survival are fractionally better.

I am transfixed, instantly in love with these exquisite creatures. Delight and awe are evenly mixed: delight at their elegance as they swim in family groups in this confined space, and awe at the toughness that has got them, albeit in limited numbers, to this point. I am overwhelmed, too, by sensations of connection to an ancient primaeval world, and of the fragility of existence. From discussion with Adam I know the threats they face, and that their survival depends on people, on science. I juggle conflicting questions: will they survive, and, if they don't, does it matter?

That encounter with the red-finned blue-eye etched itself into my memory, because something shifted in my brain. I saw the beginning of time stretching up to today, and I was on that line. What I knew about evolution, environmental conservation and ecological interdependence crystallised in one sharp, visceral jolt. My questions settled deep within me as I travelled with my husband, Peter, to remote places in Australia to discover the Bush Heritage story, sharpening my appreciation of all that we saw and heard, provoking both fears and hope for the future.

Australian Bush Heritage Fund was conceived in 1991 by Bob Brown to buy private property or lease crown land with high conservation values to protect it forever from tree-felling, overgrazing, mining and other development. Although originally called Australian Bush Heritage Fund, the name was changed later, but the organisation will be referred to throughout simply as 'Bush Heritage', the name by which it is best known. The idea gained traction because of the urgent need to protect Tasmanian forests, and in Victoria debate about cattle grazing in the high country was escalating. Brown envisaged a not-for-profit philanthropic organisation that would protect the native bushland and all the species therein for future generations, and the idea was taken up by a group of friends and supporters in Tasmania. The word spread,

PREVIOUS SPREAD Silhouette at dusk, Boolcoomatta Reserve, SA

Photo: Wayne Lawler/Ecopix

LEFT Springs at Edgbaston Reserve, Qld

Photo: Annette Ruzicka

so that by 2014, when I began working on the project, Bush Heritage owned, or had a managing interest in, at least forty properties, totalling over 4.5 million hectares of Australian bush. I visited most of these properties, and met Bush Heritage employees, Board members, donors, volunteers, contractors, former property owners, other landholders, and people who came to live or have lived their life in the outback. To everyone I posed variations of the questions that had overwhelmed me beside the Edgbaston spring: Why should we conserve and protect our bush? What will happen if we don't? Their responses brought together different strands of the history and gave me the structure for this narrative. They also showed me the soul of the organisation.

For, in uncovering the history of Bush Heritage, I came to see that this is essentially a story about people – their journeys, achievements and failures – and about passion and commitment to a vision. I saw Bush Heritage staff, both in the Support Centre and out in the back blocks, working on reserves as pioneers, on the frontier of a different way of dealing with the bush that still seeks general public acceptance. Don Watson, in *The Bush: Travels in the heart of Australia*, draws on the memory of his grandfather's property in Gippsland in south-eastern Victoria and earlier generations of farmers to evoke the hard work, the deprivation, and the 'making do' that consumed the life of the early settlers, as they felled trees and cleared land.[1] They were nurtured by their communities, and the knowledge that they were doing essential work gave them an edge over city folk. Those living on Bush Heritage reserves today funnel their energy into restoring vegetation and conserving remnant bushland. They manage huge tracts of land solo, and must develop their own networks, because local communities are disintegrating.

The survival of the red-finned blue-eye is a rallying point for Bush Heritage that weighs heavily. It could be seen as the raison d'être for everything the organisation is attempting to do, or, alternatively, as a distraction from its core objectives, the drag on finances to protect one very rare species distorting the long-term picture. For conservation is now about big landscapes and the creation of buffers that will preserve the breadth of biodiversity against invasive plant species and foreign predators, and protect against unnatural weather events, soil loss, climate change, creeping salinity and inappropriate development, because it is recognised that the survival of the human race depends on the continuing good health of the land, the inland waters and the oceans.

While nothing is resolved, it indicates the breadth of issues facing the organisation, a yardstick as to how far it has come since Bob Brown invited friends to meet on his property. They came to support an impulsive gesture, proposed a vision for Australia, and kick-started the enterprise with an office and a manager. Thus was the Australian Bush Heritage Fund conceived under the shade of a lone walnut tree on a hillside above the Liffey River.

But Bush Heritage did not just come out of thin air, nor was the support it garnered accidental. The cultural, social and political mood was ripe for such an organisation, and its ideals and aspirations struck a chord in the national psyche. Bob Brown's political career has been thoroughly documented, but less well known is the story of Bush Heritage Australia, the pragmatic expression of his philosophy. Perhaps both began as a result of a conversation Brown had with Paul Smith on the steps of the public library in Launceston.

Author Sarah Martin getting acquainted with the red-finned blue-eye, Edgbaston Reserve, Qld

Photo: Peter Martin

PART 1

The Australian Bush Heritage Fund – as it was then known – was founded in 1991 on a grand vision, and was driven by people with passion and commitment. Its initial success had its roots in the worldwide environmental movement that had been gathering pace for decades, and in the Australian political climate at that time.

Bob Brown's vision

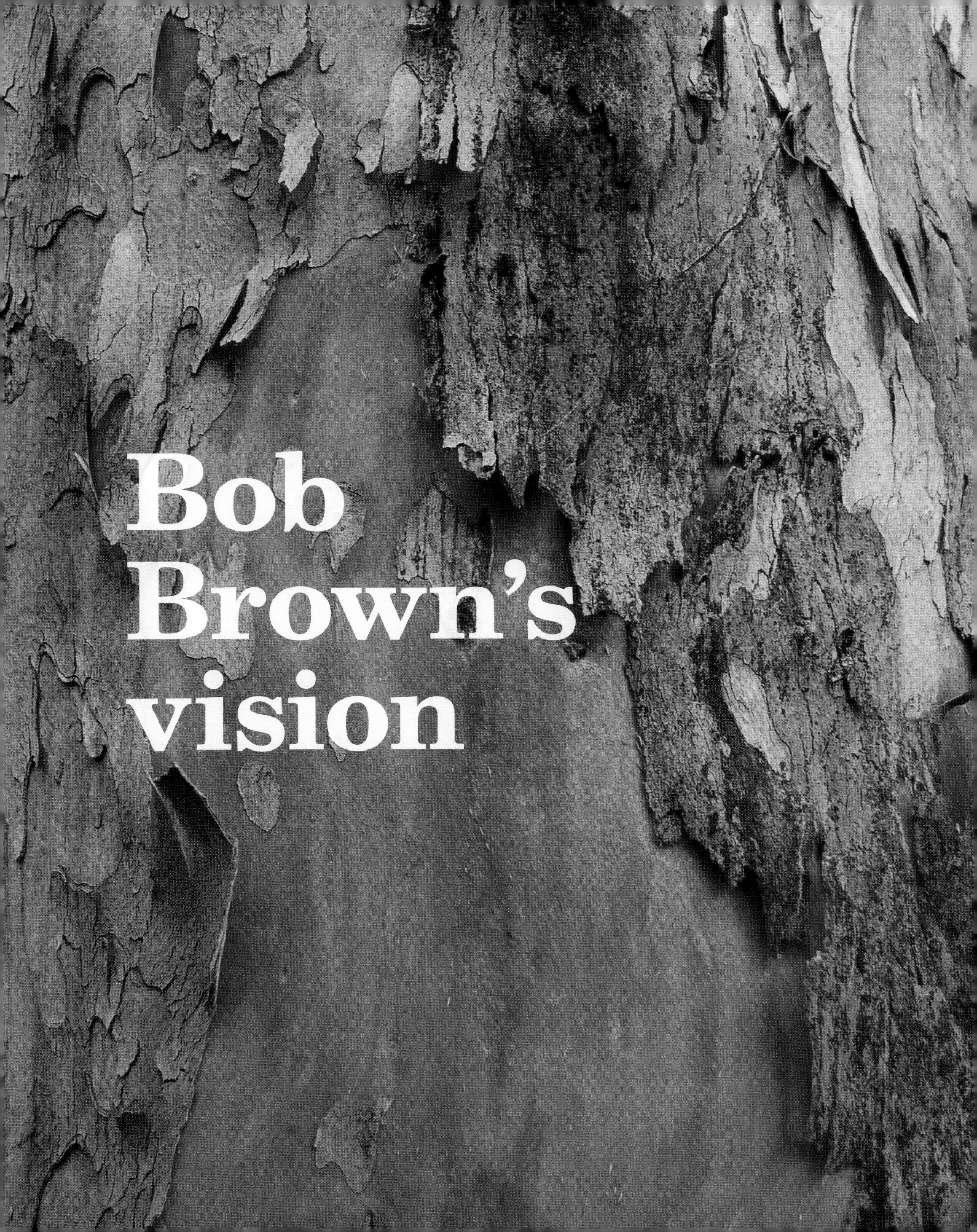

Making a mark on the environment

In 1976 Paul Smith, an experienced forester, was planning an exploratory boat trip down the Franklin River in mid-northern Tasmania. He needed a companion. A keen bushwalking friend, Peter Fleming, suggested he 'should talk to Bob Brown. He's going to make his mark on the environment'.[1]

BROWN, THEN THIRTY-ONE, had been a doctor in a Launceston medical clinic since 1972. A month before his arrival, Dr Richard Jones established the world's first Green party, the United Tasmanian Group (UTG), in Hobart in the wake of the flooding of Lake Pedder, subsumed to feed the hydro-electricity scheme. Hydro-electricity, along with logging, accounted for most of Tasmania's economy and jobs. Both industries depended on the destruction of millions of hectares of Tasmania's unique 'wilderness' areas. Brown was sympathetic to the ecological principles that UTG espoused, and he joined the radical group that was still reeling from its fight to save Lake Pedder. Three years later, Brown stood with Jones on the party's Senate ticket for a seat in the Australian Parliament, without success.

Brown had found his peace in Tasmania, revelling in the wild landscape, pristine forests and peaceful valleys. He sent a postcard to his parents in New South Wales on which he wrote: 'I am home'.[2] Soon after, he found a simple weatherboard cottage in the Liffey Valley that would indeed be his home for the next forty years.

The Liffey Valley, 50 kilometres west of Launceston, narrows as it winds its way south-west. The lower banks and flats of the river are patterned with a mosaic of sleepy farms and paddocks tucked between stands of tall trees. Thick undergrowth dips from more lofty heights down to the water's edge, curving upward on one side to meet the collection of bluffs known as the Great Western Tiers. The dolerite rock cliffs of Drys Bluff, the highest peak at the western end of this range, rear 1000 metres up, well above the tree line, directly behind Brown's cottage on the cleared hillside, forming a dramatic backdrop. The cottage, built in the early 1900s, has three steps up to the verandah overlooking the front paddock, and a pitched red corrugated iron roof. The river follows the contours of the hill across the bottom of the paddock, and the only access is via the small footbridge. While the cottage is visible from some points of the road when arriving by car, the road with its infrequent traffic makes no impact once the bridge is crossed. Brown paid $8000 for the cottage and 10 hectares of 'vibrant forest and meadow, [with] nearly a kilometre of riverbank ... the barred bandicoots, rare sweet-faced bettongs, ruby-eyed goshawks and boobook owls were included gratis'.[3] He called it Oura Oura, and it became his place of 'intense contentment', his refuge from the world of environmental activism and politics that drew him away from his medical career.

Full of risks and rare experiences, the trip was life-changing. Brown never forgot the raw magic of the river.

From Oura Oura he articulated the ideas on the environment and democracy that underpin his first book, *Memo for a Saner World*, first published in 2004. Smith remembers late-night discussions with Brown on philosophy and the environment while Brown staffed the Launceston medical clinic during the weekends. On Monday mornings he would cycle back to Oura Oura to resume his writing.

Smith was concerned that the majestic forests and sinuous rocky contours of the Franklin River were about to be destroyed. Although it hadn't been announced yet, he knew that the Tasmanian Hydro-Electric Commission was already working on the damming of the Gordon River into which the Franklin River flowed, and heavy machinery had appeared in the Franklin River area. His bushwalking friends would not join him on the trip down the Franklin, judging it too dangerous. But Brown was always ready for a challenge. He had never walked into Lake Pedder with its glorious sandy pink shore, deep in Tasmania's wilderness, but knew he didn't want any more bush to be destroyed. A chance meeting at the library encouraged Smith, on impulse, to suggest the trip, and Brown readily agreed.

Although both men were fit bushwalkers, neither had experience on water. They considered several options before deciding on inflatable rubber boats. Smith constructed long, double-bladed paddles with a prong extension to push off from rocks, which they tried out in a local dam, but they were useless, spinning them around in circles. Through trial and error, they found that by sitting in the back, bolstered by lilos and plastic cushions, their legs either side of their packs, they could propel themselves forward while still maintaining balance. They travelled light, in contrast to an earlier expedition in the 1950s by John Hawkins, which had involved vast quantities of supplies in heavy wooden Canadian canoes. Hawkins told Smith and Brown that the river should be dammed to stop people risking their lives, since it was one of the wildest rivers in the world. Undeterred, Smith and Brown set off.

Full of risks and rare experiences, the trip was life-changing. Brown never forgot the raw magic of the river, the peace of the wild country, and the shock of discovering that the Tasmanian Hydro-Electric Commission was ready, with drills and dynamite, to flood the area. Invigorated by their adventure, they returned the following year with friends to make a short film. *The Last Wild River*, by Smith and Amanda Stark, was a potent weapon when, in 1979, the Hydro-Electric Commission of Tasmania announced it would dam the Gordon River, an action which would impact on the Franklin River.[4] The Wilderness Society (TWS) had been hastily convened from the remnants of the Tasmanian environmental movement, still in disarray after the loss of Lake Pedder, and in 1978 Brown became its Director in preparation for a more organised, better financed battle. TWS was joined by the Australian Conservation Foundation (ACF) and the Tasmanian Conservation Trust (TCT), with Brown spearheading a nationwide, multi-layered campaign to stop the dams.

FOLLOWING SPREAD **Bob Brown's cottage, nestled between the Great Western Tiers and the Liffey River**

Photo: Peter Morris

In 1980 over 10 000 people marched in Hobart demanding the government not proceed. Brown travelled Australia, speaking against the destruction of the environment, and bright yellow triangular 'No Dams' stickers appeared, along with the hauntingly beautiful photographs of Franklin River wilderness areas taken by Peter Dombrovskis. Dombrovskis's dramatic photo, *Morning Mist, Rock Island Bend, Franklin River*, was captioned with a question that caught the emotion of the moment – 'Would you vote for people who want to destroy this?'

The campaign generated international interest and over 30 000 letters of support. Hundreds of volunteers arrived in Tasmania to help in whatever capacity was needed. But all would have failed had not three things occurred: Bob Hawke, leader of the Labor Party, and always attentive to the electoral mood, promised that, if he were elected, no more dams would be built in Tasmania;

Morning Mist, Rock Island Bend, Franklin River, Tasmania 1979

Photo: Peter C Dombrovskis, © Liz Dombrovskis

in 1982, the World Heritage Commission in Paris voted to list Tasmania's wilderness and wild rivers, including the Franklin, on the register of World Heritage properties; on the same day, Brown began a peaceful blockade to stop the machinery. During the blockade 1300 people were arrested, and 500 gaoled, including Brown, who spent nineteen days in Tasmania's maximum security Risdon Prison. Images of peaceful protesters being violently manhandled appeared on television screens throughout Australia, swaying public opinion in support of the 'No Dams' campaign.

Hawke won the 1983 election in March, and passed the necessary legislation to stop the dams. In July the High Court knocked back Tasmania's Premier Gray's challenge to the federal government's right to protect its Tasmanian World Heritage property. The Franklin River protest lasted over six years, ending in personal triumph for those who fought the battle. Its aftermath was bitter: the stand-off between the protesters, and the Tasmanian government and its pro-dam supporters, split the state and stirred animosity that still festers today, while the fight to stop the loggers destroying other forests continued. Brown, with his thin, willowy frame and steadfast, polite demeanour, was the iconic figure, the hero who saved a pristine wilderness. His calmness, even when roughly bundled into a paddy wagon by two policemen, was fixed in the national memory. He seemed to have singlehandedly brought the federal government to its senses in time to save the river. A simplistic picture, telling only part of the story, as Phillip Toyne, ACF Director at the time, indicated in wry but admiring comments in 2010. Despite the Conservation Foundation's huge contribution to the campaign, no one remembers the ACF; they only remember Bob Brown.[5] Peter Fleming's prediction that Brown would make his mark on the environment had come true.

For Brown, the Franklin River campaign was the catalyst for a change of direction that manifested in two different but connected ways. Firstly, it reminded him that he was a conservative person at heart, with little appetite for activist campaigning and the violence associated with protest. He did not dwell on negatives, preferring practical ways to achieve positive outcomes. Politics had some appeal, and suddenly became even more attractive when Norm Sanders resigned from the Tasmanian House of Assembly and the seat was offered to him. As Brown remembers: 'We were locked up in our single cells for fourteen hours every night, so I had plenty of time to ponder taking up the vacant seat of Denison. In the event, I took the seat and was elected in the subsequent countback for the seat the day after I was released from prison'.[6]

The Franklin River protest lasted over six years, ending in personal triumph for those who fought the battle.

Connections between Sanders and Brown were strong. Sanders, a scientist and a member of the Democrats Party, had been the inaugural Director of TWS. He had written groundbreaking critiques of the environment and made radical (at the time) suggestions for fire management in forests. He was the first politician in Australia to win a seat on an environmental platform alone. He had resigned over the government's handling of the Franklin Dam and the treatment of the protesters. Following the trailblazing Sanders was the start of a distinguished career in politics for Brown as the leader of the Greens, first in the Tasmanian parliament, and later in the federal parliament, where for a time he held the balance of power, using it to advance environmental causes and human rights. Those first years in politics that began in 1983 took up most of Brown's time and energy, and he later commented that 'the parliament was a far less friendly place than the prison'.[7] It was not until 1990, when he won the Goldman Environmental Prize for his work on the Franklin River campaign, that the environment could once again take centre stage for him. The Goldman

FOLLOWING SPREAD **Drys Bluff Reserve at the foot of the Great Western Tiers, one of the first Bush Heritage reserves**
Photo: Wayne Lawler/Ecopix

Prize would be the catalyst that again changed Brown's direction, and ultimately led to the birth of Bush Heritage.

San Francisco couple Richard and Rhoda Goldman's philanthropic foundation had given away over half a billion dollars since 1951. By 1989 they began focusing on the environment, and established a prize, sometimes known as the 'Green Nobels', to honour environmentalists who were grassroots activists, one person from each continent, ordinary people who were capable of doing truly extraordinary things, recognising 'individuals for sustained and significant efforts to protect and enhance the natural environment, often at great personal risk'.[8] The prize was inaugurated in 1990 and Brown was the winner for the continental region of Islands and Island Nations. Although 'not one for prizes', Brown decided to go to San Francisco to receive the prize, because he

Liffey River Reserve, Tas

Photo: Wayne Lawler/Ecopix

wanted to present President George HW Bush, whom he was advised he would meet, with a letter appealing to him to protect Antarctica as a world park, and because he felt he 'owed it to all the thousands of people who had worked on the Franklin River campaign'.[9] The prize was a cheque for A$49 000.

On his return to Australia in April 1990, he discovered that two privately owned Liffey bush blocks were listed for auction. The blocks were adjacent to, but outside, the protection of the Drys Bluff and Liffey Falls Forest reserves. Loggers in the early twentieth century had been through the area with axes and crosscut saws, selectively milling the timber in sawmills within the forest. Now, the blocks were vulnerable to another round of logging and, once logged, would open the way to further tree-felling in the wilderness on the other side of Drys Bluff. Brown knew that the loggers in the 1990s would clearfell the forest, using bulldozers, chainsaws and cables. The blocks were worth $240 000, and Brown realised that 'fortuitously I had in my pocket the prize money which was enough for the deposit'.[10] On 29 May he rang his good friend and fellow environmentalist, Judy Henderson, in Oxford, to tell her that he was going to bid on the blocks. Henderson advised him not to buy them, because she knew he had no money to back up the purchase.

Three days later Brown rang her again, admitting he had bought the blocks for $240 000, using the Goldman Prize money as a deposit. Henderson flew home and they discussed a course of action. Brown's impulsive gesture had landed him with a debt of $200 000 and no means of repaying it. They had to devise a plan to pay for the blocks within 130 days or the deal would fall through. The only thing to do was to obtain a loan from the bank and then raise the money to pay it off.

Securing such a loan was not easy. Brown suggested that the bank use the blocks as collateral since he had no other sources of income. Ironically, should he fail to pay back the loan, the land would be cleared for woodchips, valued at $400 000, by North Broken Hill. The first bank manager they approached was uncomfortable with Brown's proposition because there was no monetary return from the land to service the investment. However, the second bank manager was more sympathetic and wanted the loan to go through, although he must have known it was questionable, as Henderson discovered later. In 1992 the loan was taken over by August Investments, an ethical investment company which became Australian Ethical Investment Ltd (AEI). When Henderson became the Chair of AEI, her understanding of board decisions made it clear that AEI would never take on such a risky investment again.[11]

Brown's impulsive gesture had landed him with a debt of $200 000 and no means of repaying it. They had to devise a plan to pay for the blocks within 130 days.

But it was never going to be a simple loan repayment. No one now remembers when and how paying back the loan morphed into the idea of establishing a fund to buy other blocks of land, but a plan emerged at the first meeting at Oura Oura of a group of Brown's friends who were keen to help him out. Brown already knew about The Nature Conservancy (TNC). As the biggest philanthropic environmental fund in the United States, with an annual budget of US$500 million, TNC had been buying up land since 1950 to save it from future exploitation, and to manage it into perpetuity. Brown had made a reference to TNC in an article he wrote for the environmental magazine *Habitat Australia* before he bought the Liffey bush blocks.[12] Clearly, even before he bought them, he knew he would have to raise money to pay for them. It took just one imaginative leap to decide to emulate the American example, and Brown admits that he and Henderson had flirted with the idea before the meeting.[13]

The friends who met at Oura Oura liked the idea, and began planning. Believing that names should express purpose, they decided on 'The Australian Bush Heritage Fund', to indicate from the outset its intended reach. They established a small fundraising committee, and developed a brochure explaining the Fund's aim. Peg Putt, one of the group, remembers the brochure's attention-grabbing headline: 'Suddenly the chainsaws were coming'. The threat was clearly outlined, with a plan to raise money to buy the forests and save them from the chainsaws. 'Somehow we got it inserted into the ACF's *Habitat* magazine in October, a real coup, as we never got near the magazine again, and we got a huge response.'[14]

Having secured the loan, they now had to work out how to pay it off. But events in the Tasmanian parliament overtook this discussion. The Labor–Green Accord that Brown and his four Green colleagues had secured with Premier Michael Field, enabling him to form government after the 1989 election, was showing signs of the stress that led in 1992 to its collapse and another state election. It was not until September 1990 that Brown and Henderson met with friends Dave Heatley, treasurer of the TCT, Sue Rundle, a medical researcher, and Ian Skinner, a fellow environmentalist and graphic designer, to brainstorm on fundraising. From there, as Brown said, 'We just went from one gear to another'.[15] They drafted a letter requesting support, and an advertisement for the newspaper, and Skinner designed a logo of three trees, the distinctive icon that first created Bush Heritage 'brand' awareness. TCT and Bush Heritage established the appeal through a fundraising vehicle under TCT's auspices, with the catchy title 'Buy Back the Bush'. All funds would be channelled through TCT, which had tax deductibility status, and Margaret Atkinson, a part-time employee of TCT, would deal with incoming mail and cheques. With all this in place, the Australian Bush Heritage Fund was formally launched on the Liffey block on 26 November 1990, attended by Channel 9 and ABC local radio's Liz James, who became a long-time supporter. Tasmania's newspaper, the *Mercury*, carried a small article about the launch the next day.

Brown's article in *Habitat* about the Liffey blocks had already generated enthusiasm; donations of $10,

BELOW The logo of three green trees illustrated the focus of Bush Heritage's efforts to protect the bush.

RIGHT The success of 'Suddenly the chainsaws were coming' set high expectations for future campaigns.

Source: Bush Heritage Australia

AUSTRALIAN BUSH HERITAGE FUND

$25 and $50 from people all over Australia came with personal letters thanking Brown, and pleading for other pieces of land to be saved. Clearly, the idea had broad appeal. Generous loans, including $50 000 from landscape painter Graeme Cox, were valuable, but not enough to ensure survival for the blocks. The loans carried their own conditions: some were interest-free, others specified the amount of interest required and whether it would be payable to Bush Heritage or to the lender, and there were varying lengths of time before the loans had to be paid back. The accounting was complicated. Later on, letters ranging from polite enquiries about when the money was going to be paid back to 'I want my money back' added to Bush Heritage's early administrative problems.

Brown called a friend with experience in fundraising and marketing. Verna Simpson, currently Director of the Humane Society International, has always been a committed environmentalist, and a 'sucker for animals in plight'.[16] On her return to Australia after time working overseas in fundraising, she helped establish Greenpeace Australia and the Australian Fund for Animals. She also organised what was probably the first environmental protest march in Australia, which saw protesters don gasmasks to walk across Sydney Harbour Bridge to highlight the problems of air pollution. Later she purchased the magazine *Simply Living*, turning it into an environmental journal, giving non-governmental organisations (NGOs) a regular page in its 'Green Pages' section, so that it became the 'go-to' magazine for the environment. She maintains that fundraising for the Australian Bush Heritage Fund was the easiest fundraising she had ever done, because the story captured people's imagination. 'Bob kicked off the purchase with his $49 000 prize money from the Goldman award, sending a powerful message emphasising the importance of the project. He had also bought the properties without the money to back up the purchase, so it was a desperate plea for help.'[17]

The gesture inspired many, including Tim Mulligan, who wrote to Brown offering him an interest-free loan of several thousand dollars, to be rolled over to Bush Heritage when it became operational.[18] He and Brown had not met, but the financial arrangement was made and agreed upon through letters, with honour as the unspoken security. Other donors also came forward to help Brown with his mortgage, and by the end of 1990 Bush Heritage had banked another $11 500 from seven donors. But most of this was in the form of loans that would need to be covered by donations. The early financial records, painstakingly handwritten by Brown, with detailed explanatory notes, indicate the precarious and amateur base from which the organisation grew.

Donations of $10, $25 and $50 from people all over Australia came with personal letters thanking Brown, and pleading for other pieces of land to be saved.

The fledgling operation began with a desk in the office of TCT. A storeroom, with bars on the window, was cleared out in a small office in an old sandstone building in Bathurst Street, Hobart. The office opened onto the communal kitchen shared by TCT and the Environment Centre, and was host to a constant stream of colourful visitors – staff, volunteers and others who dropped in to make themselves a drink, retrieve their lunch from the fridge, or dry wet socks. Those first Board members joked that Bush Heritage began its life in a cupboard so small that one could touch every corner of the office without leaving the only chair. Atkinson was inundated with new donors: as Putt commented, 'It was a time that had come. People wanted to do something positive and see results. While the political work on conservation had been successful, you never knew whether it would last'.[19] The initial response exceeded all expectations.

Oura Oura's warm welcome

On a brilliantly sunny early morning in February I drive from Launceston to Oura Oura to meet Bob Brown. The valley sparkles green and lush, and a cloudless blue sky picks out the ragged line of the bluffs on my left. I search for the red roof that Paul, Bob's partner, has told me to aim for. I miss it, and have to backtrack, feeling stupid, arriving late.

I needn't have worried. Bob is pretty laid-back, as everyone who knows him will tell you. He has a calm, engaging voice that spikes frequently with his dry wit and a subtle turn of phrase which, if you are not careful, you might take at face value. He and Paul make tea in a billy on the wood fire in the tiny living room, almost like camping, and we chat. We are still chatting over a tour of the cottage, very short because the cottage is even smaller than I had imagined, and as we wander up the property to the fence line and the gate from where the almost vertical walk to Drys Bluff begins. It is not mindless chat, though: Bob is bringing me into his world, his 'anchor in the drift' that is Oura Oura, casually naming each species of tree as we pass, pointing out delicate plants among the leaf litter, identifying bird calls that float down around us. He tells me that in 1992 fern expert Michael Garrett discovered that Drys Bluff was home to *Asplenium hookerianum*, Tasmania's rarest fern, previously known to grow only at Hellyer Gorge on the edge of the Tarkine Wilderness and another remote site in south-east Tasmania. Breaks in the tall green canopy show a wedge-tailed eagle lazily circling above us. Higher up is the guano-stained ledge on which peregrine falcons have made their nest, and Bob describes the moment when two baby falcons recently took to the sky for the first time. I am beguiled by his stories and the matter-of-fact information about the environment he knows and loves, and I feel the passion.

During lunch, on the concrete strip behind the house, lizards dart in and out of the stone wall. But it is the other wall, the jagged rock fortress of Drys Bluff, so close above us, that dominates and mesmerises. I will come back and climb that peak one day.

After lunch we drive the short distance to the first of the Liffey blocks, where last century loggers set up a small sawmill. Remnant stumps nearly 2 metres in diameter indicate what has gone forever. But the forest we walk through is still awe-inspiring. We cross the creek that ripples over sun-splashed rocks, walk through a cool and secret place of tree ferns and mossy roots which, as we climb, gradually gives way to stands of smooth white straight-limbed eucalypts among tall grasses, slimly graceful against the blue sky. Later, as we clamber our way up the hillside of the other block, slipping on shiny eucalyptus leaves and gravelly stones, and stepping over fallen trees and hollow logs that make perfect habitats for lizards, snakes and small furling fans of orange fungi, I am drawn deeper into Bob's world.

Two couples arrive during the afternoon to swim and wander around Oura Oura, taking advantage of Bob's well-known 'open gate' policy that gives everyone the right to explore this magical place. Though the property is now owned by Bush Heritage, Bob's sign – 'Trespassers welcome' – still remains on the bridge. The visitors are fervent, admiring and grateful, all of which Bob deflects with dismissive kindness, while Paul, with a wicked grin, winks his amusement. As Bob explains later in the mildly satirical way he describes his own actions, 'Half of Tasmania may think I am a saint, but the other half thinks I am the devil incarnate!'

I swim in the rock pool beside the footbridge where the afternoon sun lingers. Floating in the crystal water, I search the bank for the platypus that Bob tells me lives here, but the shy creature, not surprisingly, refuses to show itself. After my swim I stretch out on a warm rock beside the pool to soak up the last of the mellow afternoon. As the sun slides away, we have dinner on the slope above the cottage overlooking the walnut tree. Leaping flames from the wood fire form a shield against the cool air stealing up from the river and down the mountainside. A lone star pricks the sky and the half-moon lifts higher above the low hills on the other side of the river, devouring the glowing pink twilight. The peace of Oura Oura has cast its spell, and in a dreamlike state I take the road back to Launceston.

Bob Brown's message to 'trespassers' – still in place in 2016

Photo: Peter Morris

Harnessing the chaos and energy

For those who took part in them, the first years at Bush Heritage were 'exciting, wonderfully exhilarating, and stressful, incredibly stressful'.[20] If the 'cupboard' in the TCT office in Hobart was the engine room of Bush Heritage Australia in the early days, Oura Oura remained its spiritual centre.

FOR THOSE WHO WERE THERE, the small house in the clearing enclosed by tall trees, unpathed bushland and craggy cliffs, symbolised the triumphs and the difficulties of the first few years. The organisation took its first steps, faltered, and could have crashed, but was held together and steered onto a secure footing by the group's belief in its possibilities. That summer the scent of sun on leaves and grasses, rain falling on damp earth, and the clicks, twitters, rustles and buzzing in the bush were justification for every ambitious idea generated from passion-filled meetings under the lone walnut tree. In March 2016 huge winds brought down the tree. Brown was devastated. He set about growing a new tree from saved walnuts, and commissioning the carving of a beautiful object from the timber to keep the memory of that inspirational time alive. But miraculously in May, after heavy rain, the fallen tree began to sprout again, matching the resilience of the organisation it had first nurtured beneath its generous canopy.

Like any group starting up a business, there was an oversupply of ideas – some good and others too wild to contemplate – little knowledge of the systems needed to carry them through, and never enough money. The Board members, professionals in their own fields, came together as amateur enthusiasts, with only sketchy ideas about developing an organisation from scratch, starting with an unwieldy structure and ambitious agenda driven by the desire to create a national organisation. Brown remembers the first two years as 'hard, very hard. We were amateur but passionate, believing this was the right thing at the right time, and that it would fly if we got the instructions right. We knew we had to develop good management to stave off imploding before we even got off the ground'.[21] They were 'hands-on' directors, compensating over and over again for an almost non-existent budget. Judy Henderson's used teabag pinned to the notice board symbolised that struggle. They survived, and Bush Heritage flourished, because of their belief and their hope. Most had already experienced disheartening aspects of the conservation movement where all they could do was hold the line, or engineer a managed retreat. Bush Heritage offered a constructive way without the usual bitter struggle and conflict – it was 'soul ease' for 'battered warriors' of the conservation movement.

Although ambitious, the Bush Heritage message was simple and attractive: buy land and protect it to allow native vegetation and species to flourish. While Brown was the inspiration and could develop big-picture visions, he had also just become a Green Independent in the Tasmanian parliament, which meant that he left most of the work to others, confident of their ability. As a leadership style, it was brilliant in fostering a committed community, and empowering others to take control, and to 'own' the project.

Judy Henderson, who possessed great organisational skills, took on the day-to-day business of the fledgling organisation. The first steps were to pull together a management committee (which during the early years was, of course, the Board), to find office space, and to draw up a constitution. The constitution was a top priority, enabling Bush Heritage to become a registered body by October 1991, just over a year after its inauguration. Brown remembers that 'Judy's skills and manifold contributions from the very outset at Liffey were central to the organisation's success and to its prospects

Riffles in the Liffey River as it flows through temperate rainforest

Photo: Wayne Lawler/Ecopix

in the future'.[22] She was the calming influence in times of chaos, when drafts for newsletters, plans for conferences, ideas about sponsorship, memos on the constitution, and incoming mail with cheques and outgoing mail containing receipts and letters of thanks existed in one muddled pile, and the directors disagreed on the way forward. Everyone agreed that she was 'amazing', one commenting that she 'stripped away all the bullshit and got to the point with brutal frankness but also with absolute courtesy'.[23]

The original Board comprised directors Bob Brown and Peg Putt, Director of the TCT; treasurer Alistair Graham, Putt's partner and an environmental consultant; secretary Penny King, a veterinary surgeon; and Judy Henderson, who would in 1995 become Chair of Oxfam International. Other members were Sue Rundle, fundraising consultant Verna Simpson, and Director of Community Aid Abroad's Australian operations, Jeremy Hobbs, all friends through their professional lives and their commitment to environmental causes in Tasmania. While they aimed to have monthly meetings, they were often irregular, since all were working full time. But by August 1992 their confidence that the organisation could survive prompted them to invite Brown to become President, and he took up the position at the next meeting.

Sally Port, a former teacher, who had run both a community organisation for the disabled in Sydney, and the Sydney Amnesty International office, was the organisation's first employee. Employed as a trainee fundraiser, she soon realised that her job was to set up the infrastructure for the new organisation, and she was grateful for Verna Simpson's mentoring, and the relational database Simpson established.[24] At the time, Port, Simpson, Putt and King were all young mothers, united as they juggled parenting with the stresses of work in an office with one chair and a desk, and so little room that a person having a conversation with the incumbent of the chair had to pull up a chair from the office next door, or stand in the doorway. Port revelled in the new experience and Simpson also describes those early days as exhilarating: 'I would sketch a marketing plan, design brochures and Sally did all the running about. I spent many hours breastfeeding to an audience in the small cramped space, but it all felt so natural and right, just part of our landscape'.[25]

After the success of the 'chainsaws were coming' brochure inserted in the ACF's *Habitat* magazine, fundraising became local, and Board members themselves placed brochures wherever they could: on reception desks in hotels, in cafes, restaurants and libraries, leaving a trail across Hobart and Launceston and in their home states. Once Simpson set up her direct marketing strategy, the pace quickened. The key to its success was attracting donors already engaged in wilderness issues through the insertion of brochures in magazines that catered to bushwalkers, birdwatchers, cyclists, bush regenerators and other such groups. By pulling together fifty or so small organisations, Simpson built a print run of over 100 000, a cost-effective operation spreading the word across the continent, a method that served Bush Heritage well for many years. They also contacted individuals whose interests in conservation were well known. A fax from 1990 attests to Canadian author and committed environmentalist David Suzuki's regret at not being able to donate, since he was establishing his own foundation to fund 'an institute that is seeking radical solutions that could quickly transform society'. He offered to help out with a lecture in the future, and hoped that Brown might become an honorary member of his international board once it was set up.

Bush Heritage's profile was bolstered by creating a group of patrons. Port contacted naturalist David Attenborough, but was out of the office when he rang to decline their invitation to be a patron. He wrote to her personally, however, explaining that Australia was so far away that he could not give Bush Heritage the attention it deserved. The point was not lost, and the directors drew up a list of well-known Australians who might be proactive in spreading the word about the organisation and supporting fundraising activities. Positive responses came from broadcaster and writer Phillip Adams, fashion designer Jenny Kee, pianist Roger Woodward, poet and activist Oodgeroo Noonuccal (Kath Walker), politician and peace activist Jo Vallentine, and singer-songwriter John Williamson, with politician Tim Fischer a later addition. Over the years Adams has interviewed Brown and other Bush Heritage personnel on his ABC Radio National program *Late Night Live*, to highlight particular projects; Woodward offered to direct a special Mozart concert, featuring the Tasmanian Symphony Orchestra and visiting Italian pianist Claudio Arrau, as a fundraiser. When the concert was cancelled at the last minute, he subsequently played for an audience in the Hobart home of a donor to benefit Bush Heritage. Williamson promoted Bush Heritage through his concerts, donating some of the proceeds to the Fund, and he and Jenny Kee would later record community radio announcements for Bush Heritage. Port remembers the interactions with patrons as 'a lot of fun'.[26]

David Suzuki offered to help out with a lecture in the future, and hoped that Brown might become an honorary member of his international board once it was set up.

The Board also set up a Scientific Advisory Committee, admittedly symbolic initially since no meetings were held at the time, but the committee members would become a source of advice in the future about properties in their geographic regions. The purpose of the committee – 'to develop selection criteria for land purchases and to review and make recommendations on land under consideration'[27] – gave insight into future plans. Professor Jamie Kirkpatrick, from the Department of Geography and Environmental Sciences at the University of Tasmania, whose research was in eucalypts and other Australian flora, was pleased to put his name to an organisation that could make a difference to the way conservation developed in Australia. He was a mentor for environmental science student Greg Blake, who would join Bush Heritage first as a volunteer, then as a member of the Board in 1993. Blake would resign from the Board in 1995 to become Bush Heritage's first full-time employee in land management. In the 1980s, two of Kirkpatrick's students, Louise Gilfedder and Rod Fensham, had assisted him in carrying out the first systematic ecological survey of grasslands and grassy woodlands in Tasmania. Their identification of conservation priorities that were largely confined to private land would be significant for Bush Heritage, and the thread of this project would run alongside the Bush Heritage story for several years before they came together. Kirkpatrick was joined on the Scientific Advisory Committee by Margaret Robertson, Peter Hitchcock and Harry Recher.

Margaret Robertson, from a farming family in Western Australia, became involved early with environmental concerns, having watched her father clear one of her favourite bush paddocks when she was eight years old. As a teenager, she joined protests against mining in the jarrah forests in the south-west of the state, and immediately after leaving school had, with her parents' blessing, followed her brother Peter to Tasmania to join the Franklin River protest in the early 1980s. She met and became friends with Brown while working on various facets of the campaign, including arranging Brown's national tour, and he asked her to join the committee as a 'wilderness expert'. In 1992 she moved back to Perth to major in geology and anthropology before transferring to Edith Cowan University to work with Pierre Horwitz (who later joined the Bush Heritage Board), with whom she collaborated on forest litigation to stop the logging of old growth forests in Western Australia. She joined the Board of Bush Heritage in 1994, becoming a significant player in its development.

Peter Hitchcock and Harry Recher were both acknowledged scientists in the area of environmental ecosystems, acting in various capacities at federal and state government levels on ecological issues. Hitchcock, involved in rainforest management, would win the 1993 IBM Environment Award. By 1990 Recher, teaching at Edith Cowan University, had formed the view that conservation in Australia, both governmental and private,

Bob Brown continued in the role of spokesperson at events.

Source: Bush Heritage Australia

was failing because it was based on emotionalism rather than science. He saw problems with the system of reserves: they did not represent all types of vegetation, were too scattered to achieve any useful connectivity, and were poorly managed, largely due to the anthropocentric view of conservateurs, so that the system would always be at the mercy of commercial and economic possibilities. Recher provided hard-hitting advice, but it was some time before Bush Heritage was sufficiently established to act upon it.

The Board discussed how to approach corporations and how to develop sponsorship opportunities, both of which involved tricky negotiations as to what they could offer in return for support, and what they could accept within their own ethical framework, which precluded arrangements with companies whose operations were damaging the environment. Plans to design a corporate brochure went ahead, although it was probably too early to consider this option when they were still feeling their way. That they did not reach any decisions on these matters is not surprising, since both issues were to remain vexed questions for Bush Heritage over many years.

They also drew up plans for a bequest brochure; began writing and sending out a three-monthly newsletter to their growing base of supporters; worked on presentations to groups in Hobart, Melbourne and Sydney; developed information packages about Bush Heritage to explain the project; and considered approaching the richest 200 people in Australia and the 100 top-performing companies for donations. Events for donors were planned, volunteers helped in the office, and discussions about seeking pro bono work from lawyers, accountants and IT experts were hot topics at meetings since the need for more expertise in these areas soon became obvious. In its turn, August Investments showed its support for Bush Heritage by offering to waive its entry fee for individuals prepared to donate the interest accrued on their investments to Bush Heritage. Port, responsible for most of the paperwork, was simultaneously exhausted and exhilarated from the many activities, ranging from the sublime to the ridiculous, that made up her day. On any one day, for instance, she might discuss possibilities for the rock group Dire Straits to join Bob Brown in a flyover of the Tarkine Wilderness, arrange a morning tea for donors, and approach local businesses to supply a much-needed computer!

Bush Heritage was vitally dependent on Brown's high profile, and, despite his increasing role in the Tasmanian parliament, he was always available for meetings and as a speaker at events. He did not believe that Bush Heritage should be tied to a political party and, now that he was a political figure, the inherent tension between his recent activism and conservative conservation was an issue, one that perhaps inhibited some people from supporting Bush Heritage. Bush Heritage was a positive response to environmental concerns, and part of its appeal was that it had no party alliances. It stood out from other conservation groups already operating, with its uncomplicated objective of buying land to save it. Discussion at Board meetings as to whether Brown ought to cut his ties with Bush Heritage continued, but without resolution.

The Board was also dealing with requests from people throughout Australia who had land to sell, and timely and appropriate responses stretched the capacity of an organisation already under stress. Nevertheless, they were keen to buy another block of land to prove they were serious about conservation and each block had to be considered on its merits, despite there being no mechanism for this process. Underlying the desire to expand was the nagging worry about meeting deadlines on the existing loan repayments, and stretching the money to refund those who had lent money on a short-term basis. Money was coming in, but more slowly than anticipated. Spreading the word further and wider seemed the only way forward!

Bush Heritage was vitally dependent on Brown's high profile, and, despite his increasing role in the Tasmanian parliament, he was always available for meetings and as a speaker at events.

When Port left to have another baby in 1992, she was replaced by Frank Haddon. Haddon had written several children's books with environmental themes, and had experience in office management. He began by trying to improve the systems, reorganising and streamlining the processes to shore up the inadequacies, but had no 'breakthrough' ideas about progressing the fundraising side of the business. For the Board, continually anxious about the rate money was coming in, and perhaps unaware of the behind-the-scenes processes necessary to make this happen, it was soon time to make the decision to end his appointment. Penny King remembers going with Brown after the Board meeting to give Haddon the news. She was shocked, as was Sue Rundle, and they resigned from the Board, both regarding the process as badly handled.[28] However, Jeremy Hobbs, also at the meeting, felt that the Board's decision was the right one for Bush Heritage at that stage of its development.[29]

This was one of several moments in the first years when Bush Heritage could have fallen apart, but the Board was committed, and they had donors who already had an emotional and financial stake in the Liffey blocks. Brown used his considerable persuasive powers to coax King to return to finish the newsletter that was due at the printers.

She came back reluctantly, but in the end was glad to see the organisation pull through this difficult stage. She had been a volunteer in TWS since 1989, and had been approached by Brown when he was seeking more members for the Board. She remembers the work as exciting, but 'a bit hard to get a handle on …, it was at that stage still rather ephemeral, still just an idea'.[30] She stayed on as a director until 1995 when the organisation was more firmly established. Meanwhile, someone was needed to carry on the day-to-day work of the Fund.

Victoria Templeton, a friend of Peg Putt, and an experienced writer and publicist, took on Haddon's job. It was 'a bit like taking on a galloping stable-full of wild horses: a gigantic and exhilarating job'.[31] With the new title of 'National Coordinator', she threw herself into the work, bringing to it her own artistic and literary flair. In her newsletter chats she caught the lyrical quality of the Tasmanian bush and introduced a more intimate tone that drew readers with her as she explored what conservation meant to her:

> I grew up in what was still then the 'real' Bush. Places of untouched loveliness, quietly vibrant with life. The fragile strength and cosiness of a wren's nest of eggs. The grace and speed of kangaroos. The slow spiky sweetness of an echidna. … Most of my childhood's special places have gone now (how about yours?). … You don't have to wear a label to love the Bush: it's part of our culture, our history, our childhood. By the time we declared the magnificent Tasmanian tiger 'protected', it was gone. Let's not let that happen to all our 'special places in the Bush'. We can make them part of our future if we buy and protect them now.[32]

She also coined the catchy tagline – 'We don't beat about the bush, we buy it' – that Bush Heritage would carry for some years before it was seen to be inappropriate as the organisation moved towards developing partnerships with pastoral landholders, and with Aboriginal people. She developed, as Port had before her, easy relationships with donors. If she was less diligent about the paperwork, her enthusiasm and the dramatic intensity with which she depicted the bush in dire need certainly attracted new donors.

Templeton's involvement, and subsequent departure in 1994 because she could not continue working at the frenetic pace, marked the end of the first stage of Bush Heritage's development, and perhaps precipitated a rethink in office management. In the March 1993 newsletter in the column 'The Voice from the Office' she describes an average workday, no doubt using some literary licence. Even with this caveat, the article captures the essence of the work, the spirit of the organisation, the chaos and the energy that characterised those early years:

> But seriously folks,
>
> In between thinking of what to say to you, the astonishingly wonderful people from every walk of life, every profession, all over Australia (and hello overseas!), who are concerned enough about what's happening in our backyard to actually do something (non-political) to help, I'm also meeting the same deadlines for five or six full magazine articles, complete with maps and photographs, writing, designing and organising production of our donation brochures, our information literature, our corporate brochure, bequest brochure, answering correspondence, much of which requires research and materials, handling the banking, bills, legal and government requirements, booking brochure insert programs with magazines all over Australia and all the printing involved, while answering and making phone calls on every conceivable topic, getting the Latin names (of plants and animals) right, trying to find time to respond to all the people wanting to get in touch with others doing related things, planning to do the filing soon, and berating myself for not being out there promoting the Bush in the large-scale way it needs if we're going to save some while it's still there.
>
> In short, *having a wonderful time, please send money.*

A changing environmental movement

Brown was keen not to have to reinvent the wheel, and arranged to visit TNC in the US soon after Bush Heritage was incorporated. He was warmly received and met with key people who were happy to share their knowledge and experience.

THE SIZE OF TNC'S OPERATION was astounding – thirteen departments, a US$1 million entry requirement for Board members, autonomous chapters in every state, over 3000 employees, and several million donors and volunteers. TNC's thinking on land conservation had already started to move to a different level: buying land was becoming peripheral to building stakeholder alliances and science capabilities that could support seventy-five bio-reserves of high biological diversity around the world.

TNC began in 1950, when Illinois couple, George and Barbara Fell, moved to Washington to build a national organisation that would 'have the size and strength to take its rightful place in a country that does things on a gigantic scale'. George had degrees in botany and wildlife management, Barbara was a medical technician, and together they created programs to train people to work in the environment, and a fundraising arm to provide philanthropic support. They believed each state group would need autonomy to develop plans of local significance backed by its own fundraising group, while remaining accountable to the central Board, a model patterning itself on the established traditions of small nature conservation groups in the US, usually managed by local and regional communities. As 'The' Nature Conservancy, it gained a significant edge and, much later, when the internet changed communications, achieved overwhelming dominance on the search engine pages. When Brown visited for the first time, TNC was untouched by the scandals that broke in 2003, when its opaque associations with the corporate world attracted national criticism. Although on a grander scale from the beginning, it had started in much the same way as Bush Heritage, with an ambitious vision and little knowledge as to how to achieve it. Brown learnt that thousands of people in the US were already giving practical expression to their concerns with financial donations and volunteer work across the country. It increased the incentive to see something similar develop in Australia, since, from the time of the Franklin Dam protests, he knew that there was a growing desire among Australians to 'get serious' about the environment. And at the pivotal moment in the life of the new organisation, the question must be asked: where was this drive to protect nature coming from?

In retrospect it is possible to see a quickening of interest in the environment from the 1960s onwards, for while it had been recognised for centuries that the natural environment and its resources were being depleted as populations grew, ways of living were changing in the twentieth century, disturbing the life-giving forces of nature at a much faster rate than ever before.

Since the 1960s many had come to understand that the threats to favourite pieces of land or broader landscapes were multiplying, as were the losses of animal and plant species, and that this reduction in natural resources would eventually have an impact on food and water and other products essential to modern living. The presence of salinity in the topsoil, the failure of crops and pastures, strange weather events that provoked devastating droughts or floods, and threats posed by 'climate change' (a term that had its first public airing in the 1970s), put the environment firmly under the spotlight. Underlying those more concrete concerns, there was perhaps a deeper, more diffuse, feeling that the human connection with the natural environment was being eroded, that something important for the wellbeing of the human soul was disappearing. In 1854, Henry David Thoreau, author, philosopher and committed naturalist touched on it in his most famous work, *Walden*, when he wrote: 'I went to the woods because I wished to live deliberately, to front only the essential facts of life, and see if I could not learn what it had to teach, and not, when I came to die, discover that I had not lived'.[33] Much later, Simon Schama, philosopher-historian, in his 1996 *Landscape and Memory*, alluded to the 'long rich and pagan tradition that imagined forests as the primal birthplace of nations'.[34] In the early 1990s in Australia, for people who were sensitive to these issues, Bush Heritage's fresh approach gave hope that something positive was going to be done to arrest the decline in the health of the environment.

A significant moment in protecting wild places came with the establishment of national parks in the US. In 1872 President Ulysses Grant gazetted Yellowstone Park, to protect the vast system of high-altitude lakes, rivers and canyons that were home to hundreds of species of mammals, birds, fish and reptiles, grasslands and forests. And in the early twentieth century, John Muir, naturalist and conservationist, believing in Thoreau's dictum that 'in wildness is the preservation of the world', encouraged President Roosevelt to make the management of national parks a federal responsibility.[35] In 1879 Australia followed the US lead by gazetting its first major tract of land that was not to be improved, the National Royal Park in Sydney, the second such park in the world.

There was perhaps a deeper, more diffuse, feeling that the human connection with the natural environment was being eroded, that something important for the wellbeing of the human soul was disappearing.

By the 1890s there was fear in both countries that some species of animals and plants would become extinct. Crops, pastureland and towns were replacing forests, woodlands and grasslands, and imposing strains on river systems. In 1894 the New South Wales government established Ku-ring-gai Chase in the north of Sydney as a wildlife preserve so that 'marsupials and other fauna might roam and breed in safety ... perhaps the only place the future Australian will be able to see a kangaroo outside of his national coat of arms'.[36] And in 1909, a private organisation, the Wild Life Preservation Society of Australia, was established. Although parks and reserves were for recreation and enjoyment rather than for conservation, preserving the unique fauna was strong in the Australian psyche, and each state passed numerous Acts of legislation to that effect over many years until responsibility for fauna protection came under the National Parks and Wildlife Service in 1967 in New South Wales, and later in other states and areas were renamed 'nature reserves', terminology that encompassed the whole environment. Scientists, botanists and zoologists began advocating for parks and reserves as places of research into the ecology of the environments, rather than as places of recreation.

As nature reserves proliferated in the first half of the twentieth century, discoveries in science and technology were developing apace. By the 1960s alarm bells were ringing. In America in 1962, Rachel Carson's *Silent Spring* warned that science, recognised as the saviour of humanity, was also its killer. Her claim that pesticides were damaging the ecological harmony on which human life depended hit a nerve. Many scientists of the day loudly condemned her book, since science and technology were breaking new ground in almost every field of endeavour, and improving the general quality of life. Long-time Bush Heritage supporters, Brian and Diana Snape, remember their chemistry lecturer at university pouring scorn on Carson's theory, maintaining that science could solve all environmental problems. But Carson's theories were based on sound scientific enquiry and research, and compellingly written for the average reader. Her work fed into the disillusionment with government and big corporations that fuelled the 1960s counter-culture. Concern about the environment remained as an enduring legacy of this movement, and Bush Heritage was one of its beneficiaries.

Organisations established to deal with conservation issues grappled with two major problems: the first was the short-term thinking that drove the political agenda in almost every country. This behaviour tended to ignore environmental issues because they were too hard to deal with and could not be solved in the time between elections. The second was a refusal to accept that unlimited resource consumption was untenable. Brown had read the Club of Rome's 1972 manifesto, *Limits of Growth*, which, in its efforts to seek solutions on a global scale, brought these two strands together and underlined the contradiction between unchecked material consumption and finite resources.[37] The first edition sold 12 million copies worldwide. Brown, like many others, was 'spurred on' by such writers, and this would influence his thinking from then on.

In Australia, unease about the state of the natural environment increased from the 1960s, as evidenced by the number of private organisations that sprang up in every state on the initiative of individuals who had been shaken by the loss of plant and animal life in their regions. This loss was a result of the legislation passed after the First World War, and to a lesser extent after the Second World War, which radically changed the Australian landscape. Federal and state governments responded to the need to provide jobs for returned service men and women, and to increase food production, by opening up vast tracts of land for agricultural purposes. The result was that land ill-suited for long-term farming was cleared of all natural vegetation and farmed by people with little understanding of the consequences of these actions.

In 1961 Max Bourke, a young graduate agronomist employed by the NSW Department of Agriculture, travelled the country areas of western New South Wales and Queensland to work out why the wool industry was failing. An expert on grasses, he was appalled by the prevalence of soil erosion, degraded farmland and environmental damage. Perhaps, too, his sensitivities had been heightened by his interactions with Aboriginal people while working as a jackeroo before going to university. They talked about 'country' in a way that was new to him. These two experiences formed the basis of an enduring commitment to environmental causes that would later bring Bourke into contact with Bush Heritage.

For others, an early childhood experience sparked the idea that the environment was under threat. Margaret Graham, who came to work for Bush Heritage in 1993, remembered her awakening to environmental issues:

> Our family holiday shack was on a creek in north-west Tasmania. When we first went there it was a beautiful long sweeping beach with sandhills that we used to wander over. At the age of eleven, I remember being

appalled when they started building a big industrial plant nearby. People came to the settlement and it became like a suburb during the construction stage, changing the landscape forever.[38]

In a similar way, Michael Williams never forgot the destruction of 43 000 hectares of Heytesbury forest near Cobden in Victoria in 1956, authorised by Victorian Premier Henry Bolte for dairy farming. It would make Williams a lifelong supporter of Bush Heritage.[39]

For some, a love of nature might have been quickened by the *Wild Life* radio programs of Crosbie Morrison, naturalist and journalist, which ran for over twenty years from the late 1930s on Sunday nights. One Bush Heritage donor remembers his entire family listening to Morrison's engaging stories about animals, birds and insects, an experience that imbued in him a deep interest in nature. Max Bourke continued Morrison's initiative, albeit with a different approach. In 1969, as an ABC journalist, his radio programs had much stronger environmental messages, combining items on the Australian Red Centre, Cape York, the Great Barrier Reef and the Snowy Mountains with talks by scientists, poetry and music. The program took him all over Australia, where he talked with the first workers in the ecology movement, mostly agricultural and forestry scientists, since natural resource management courses were yet to be created. The series

The consequences of land-clearing, overgrazing and erosion graphically demonstrated the need for protecting nature.

Photo: Wayne Lawler/Ecopix

aired concerns about environmental issues, as opposed to simply extolling a love of the bush.

Love of the outdoors could just as easily have been inspired by family camping holidays, where exploring bushland yielded discoveries of birds and insects, strange rock formations and tracks in the sand, where wind and rain and sun shed different lights on special places. The Boy Scouts movement, and to a lesser extent the Girl Guides, gave thousands of teenage children all over Australia an experience of camping in the bush and cooking over a fire under the stars while listening to night birds and frogs. Many found the idea of destroying trees that were several hundred years old shocking, whereas the clearing of a few hectares of bush on the outskirts of a suburb to make way for a housing development was harder to assess emotionally. Australians did not have to look far beyond their immediate neighbourhood to notice that the natural bushland was disappearing in the name of 'progress', although it would take time to understand the significance of these small changes in the landscape. The disappearance of willy wagtails, blue wrens and scarlet robins in the suburban garden caused an unease that was put into context when the bigger picture of species loss in larger landscapes was brought to public attention by science and nature programs on the radio, and a succession of nature films that began with Walt Disney's *Fantasia* and culminated in David Attenborough's endlessly fascinating nature programs on television. They drew attention to fragile ecological systems and species loss, while drawing watchers into the wonders of the natural world. British ornithologists Max Nicholson and Peter Scott had also publicised losses in bird populations, and of whole species, which caught the public imagination. In 1961 Nicholson founded the World Wildlife Fund, now known as the World Wide Fund for Nature (WWF). Its mission is 'to stop the degradation of the planet's natural environment and to build a future in which humans live in harmony with nature'.[40] Thus, primed to consider the environment and to mourn the loss of natural habitat and species, there were many who welcomed the initiatives that Bush Heritage proposed in the early 1990s.

The creation of national parks that had begun in the 1870s meant that by the 1960s there were numerous destinations to which families, bushwalkers, bird lovers and nature enthusiasts could return year after year, or discover for the first time. And because of these early experiences, individuals would lobby for other areas to be included in state registers of protected lands. The establishment of the Victorian National Parks Association (VNPA) in 1952 by John 'Ros' Garnet was inspired by Garnet's abiding love for Wilsons Promontory.[41] As a national park, 'the Prom' sparks strong emotions. People turned out in their thousands in 1996 to tell Victoria's Premier, Jeff Kennett, to take his 'Hands Off the Prom', after he announced a proposal to open it up to what was seen as inappropriate development of commercial accommodation. Karen Alexander and Doug Humann, who were to play significant roles in Bush Heritage, led that campaign through the VNPA, Humann as Executive Director, and Alexander as the campaign contractor. Their experiences at this time would have a profound effect on the way Brown's vision became a workable reality.

Passion for 'the Prom' was still there in November 2013, when protesters again came together, re-creating the original campaign with 'Hands Off our Parks', to halt a push for major commercial development in all parks, and it had similar success. The tension between conservationists, scientists, recreationalists, park agencies and governments as to the primary purpose of national parks and nature reserves had existed from the beginning, while the philosophy behind national parks had expanded to encompass the belief that private ownership of cherished land could also be an effective means of protection and conservation. This view had gained traction at the start of the 1990s, and no doubt gave impetus to the development of the more conservative environmental organisations such as Bush Heritage and Australian Wildlife Conservancy that advocated private ownership of land as the basis for environmental protections.

In 1963 Francis Ratcliffe, a senior researcher at the Council for Scientific and Industrial Research (later to become CSIRO) pushed for a meeting of scientists, businessmen and policy-makers to consider establishing an organisation that would deal solely with conservation issues. He was inspired by comments from the Duke of Edinburgh, made when he came to Australia to inaugurate the Australian branch of the Duke of Edinburgh Awards (which, incidentally, also helped to introduce the bush to urban Australians, since a component of each level of the award is a trek requiring bushwalking and camping). As a result, Ratcliffe convened several meetings at which it was decided that the organisation would concentrate on exploring the social and economic causes of environmental problems. Its mission would include working within communities to influence urgent, transformative action that might deliver lasting change to support a sutstainable environment. With that ambitious remit, the ACF was established in 1966 with a grant from the federal government, and Chief Justice Barwick became its first president. Many of those who were sympathetic to the issues ACF raised, and on which it fought campaigns, would also be interested in the alternative approach laid out by Bush Heritage, when it began to make an impact in the 1990s.

The ACF brought to public attention the environmental issues dogging the Mallee region, Queensland's rainforests, the Great Barrier Reef and Central Australia, and it developed a broad base of supporters while pursuing research, policy development, education and advocacy work. Max Bourke was present at the ACF's first meetings, as was a young Malcolm Fraser. Bourke continued to support the ACF after he finished work with the ABC and took up a job with CSIRO as communications officer, charged with making environmental management relevant to people on the land.

The 1968 introduction of the Little Desert Settlement legislation by Victoria's Premier Henry Bolte provoked an early ACF campaign. The proposal to subdivide the Little Desert for agricultural and pastoral development prompted both local and urban Victorians to consider the implications of encroaching on the natural landscape, and they passionately and stridently supported the ACF campaign. Although the Little Desert appears as dry and visually unspectacular country, it was in fact home to many newly discovered species of plants and insects, to which the public was alerted through the protest movement to save it. Numerous species unique to Australia were identified in the process, framing a new position on the fragility of natural ecosystems. After a disastrous result at the 1968 Dandenong by-election, which showed him that the Little Desert was an electoral issue, Bolte withdrew the scheme; the success of the protest was an early demonstration of what citizen power could accomplish.[42]

In a more subtle way, the achievement of voting rights for Aboriginal people in 1962 marked a turning point in Australian thinking about the environment; the spiritual dimension of the Australian landscape painted a nuanced picture. Although there had always been some who knew and acted upon their knowledge, most landholders and farmers were unaware that Aboriginal peoples had managed the land for thousands of years by methods that were foreign to the land's more recent settlers. The 1967 landmark referendum in which Australians voted 'Yes' to amend the constitution to include Aboriginal people in the census encouraged greater understanding of the connection between Aboriginal culture and the land. As well, the popularisation of Aboriginal art, depicting landscape through a different lens, raised issues about land rights and the environment, a signpost to a country beyond the urban environment in which most Australians live. It was this 'country' that would expand Bush Heritage's horizons in the twenty-first century.

Many Australians did, and do still, subconsciously believe that Australia was 'nothing but bush'[43] before the arrival of white settlers, echoing the term 'terra nullius' by which Australia was defined in British colonial times. It took ten years of legal proceedings, which led to the 1992 decision in the Mabo case, for that position to be rejected.[44] The decision would encourage conservationists all over Australia to review their approaches to protecting and restoring vegetation, and it introduced the idea that bigger landscapes could be saved if scientific resources were combined with local Aboriginal knowledge of the land. Bush Heritage would begin working towards these goals before the end of the 1990s.

The achievement of voting rights for Aboriginal people in 1962 marked a turning point in Australian thinking about the environment; the spiritual dimension of the Australian landscape painted a nuanced picture.

In 1973 Gough Whitlam, Labor Prime Minister, launched the ACF magazine *Habitat*, a vehicle for essential news on conservation matters. Its continually increasing circulation soon indicated that the environmental messages were spreading. The Labor adviser, Dr HC 'Nugget' Coombs, who put Aboriginal land rights on the agenda, was president of the ACF during 1978–79, a further prod to include Aboriginal involvement when discussing the environment. Whitlam had flagged his interest in environmental issues in his policy speech 'It's Time' of 1972. Building on US President John Kennedy's coining of the term 'national estate', Whitlam confirmed that a Labor government was committed to 'preserve and enhance the quality of the National Estate'.[45] Following the recommendation of a committee of inquiry into the environment in 1973 that found that 'uncontrolled development, economic growth and "progress" were having a detrimental effect', Whitlam legislated in 1975 for a federal government authority, the Australian Heritage Commission, to manage the country's natural and cultural heritage.[46] The Commission established a Register of the National Estate as the primary source for submissions to the World Heritage listings. After Whitlam lost office in 1975, Malcolm Fraser, Liberal Prime Minister, was keen to see the Commission continue, and appointed Bourke as its Founding Director. Bourke would soon play a key role with the Thomas Foundation, whose support would later have a major impact on Bush Heritage.

Professor David Yencken, a long-time champion for the Australian environment, joined Bourke in setting up the Commission, and Bourke became its first CEO. In supporting Fraser on initiatives begun by Whitlam, Bourke discovered 'what a greenie Malcolm actually was'.[47] They halted mining on Fraser Island and pre-empted the first move to build the Franklin Dam, and the Commission prepared nominations for World Heritage listing for Kakadu, Fraser Island, the Great Barrier Reef and Lord Howe Island, in collaboration with the National Parks and Wildlife Service, scientists and ecologists. Both Whitlam and Bourke represented Australia on the World Heritage Committee, which was based in Paris. When Hawke formed a Labor government in 1983, he appointed Whitlam as Ambassador to UNESCO and Bourke continued as Whitlam's adviser. But the Commission was heavily criticised by mining and development lobbies, and became the Australian Heritage Council under the Howard Liberal government in 2004.

In 1982, in the last days of Fraser's Liberal government, the United Nations Association of Australia sponsored an Australian 'Year of the Tree', drawing attention to the serious decline in tree cover in rural Australia. The committees established in several states to undertake tree projects and promotional activities came together to form the non-government, non-profit organisation, Greening Australia (GA). At the same time, the Department of Home Affairs and Environment sponsored a national seminar to address the problem of tree decline by implementing a national tree planting and protection program. Fraser announced the National Tree program, aimed at reversing tree decline, and GA was appointed as the non-government arm by which the program would be carried out. By 1986, under Labor Prime Minister Bob Hawke's government, GA was recognised as a catalyst organisation to bring together diverse authorities, groups, sponsors and individuals to carry out a range of projects to retain and rehabilitate native trees and vegetation.[48]

GA would later buy properties for conservation purposes, but the bulk of its work is devoted to supporting other conservation organisations with the propagation of native plants, tree planting, volunteer coordination and funding for restoration projects. In the future it would work closely with Bush Heritage on several projects.

The 1970s saw ACF develop an independent activist approach, generating publicity and sympathy for the environmental movement as it orchestrated campaigns, mobilising supporters in ways that later provoked electoral backlash. ACF's radical activism was at its height in the 1970s and early 1980s, as was that of TWS. As mentioned earlier, one of their most successful campaigns was to stop the damming of the Gordon and Franklin rivers in support of the 'No Dams' policy, which had lasting implications for Tasmania, and ultimately led to the birth of Bush Heritage Australia. Building on this success, the ACF and TWS ran campaigns against the policies that allowed the logging industry to clear-fell forests mainly for woodchips, in some of Tasmania's most remote wilderness areas at Farmhouse Creek, the gateway to the forests of the upper Picton River, and in the Lemonthyme Valley, east of Cradle Mountain. The wild, sometimes violent clashes between conservationists and those supporting clear-felling inspired thousands of Australians to protest against the destruction of these forests, and to clamour for their inclusion in World Heritage classifications.

Unexpected support came from Graham Richardson, Labor's Environment Minister after 1987. As he walked through a Tasmanian forest with Bob Brown, Richardson had an epiphany:

> So utterly convinced of his cause, Bob can be utterly convincing; his passion and sincerity are very difficult to overcome … Having been shown the awesome forests and streams he wanted protected, I wanted to become a warrior for his cause. It didn't take too long to work out that we had a perfect convergence: what was right was also popular.[49]

Richardson had caught the public mood and understood the power of the growing 'vote for the environment' for the Labor Party. GA was quick to build on environmental concerns with a proposal to plant and regenerate trees on a massive scale in the decade up to 2000. Environmental issues were becoming a reality for politicians, particularly in light of the increasing numbers of scientific studies highlighting the land degradation, salinity and erosion that were affecting Australians in every state. In 1986 Joan Kirner, Minister for Conservation, Forests and Lands in the Cain Labor government in Victoria, had, with the support of Heather Mitchell of the Victorian Farmers Federation, instituted a new program encouraging farmers and others to understand that care of the environment had everything to do with better agricultural production. They called the program 'Landcare'. And in 1989, Labor Prime Minister Hawke's environmental policy statement, *Our Country Our Future*, included the 'One billion trees' program, with funding of $47 million over ten years.[50] GA was again appointed to carry out the program, and the pledge to plant one billion trees put the plight of the environment on the front page of newspapers. People from every demographic welcomed the victories for the environment, such as the High Court's banning of logging in the Lemonthyme and other southern forests, and the instigation of a federally funded review, the Helsham Inquiry.[51] In 1989 the signing of the Labor–Green Accord, between Bob Brown as leader of the Greens and Tasmanian Premier Michael Field, gave some respite and ensured, for a short time at least, that more of Tasmania's old-growth forest was protected. These wins for the conservationists encouraged conservatives and radicals alike to consider the implications of the loss of natural bush in other areas, and the 1990 federal announcement of an Ecologically Sustainable Development program, with ACF, TWS and WWF involved, gave hope that care for the environment would be favoured over reckless development.

The years of radical activism before 1990 helped Bush Heritage. Making a stand against uranium mining in Kakadu, taking to boats to protest against nuclear testing,

Bob Brown surveys the impact of logging on the edge of Liffey River Reserve, Tas.

Photo: Judy Henderson

hunting down the whale hunters, and living in forests for months on end was exhausting, dirty work, with no assurance of a positive outcome. As Brown himself commented: 'The only win you have in the environment is the right to fight the battle again'.[52] Little wonder that activists were among the first to embrace Bush Heritage. 'We didn't have to fight against something, and it was a relief', commented Karen Alexander, who spent years campaigning against environmental destruction, and who became President of Bush Heritage in 1996.[53] She supported the advent of Bush Heritage because it was intent on pursuing conservation outcomes within the social and political structure, running parallel with state and national conservation initiatives, but as a private operator.

This approach did not resonate with radical groups who wanted sweeping societal change to combat the environmental devastation they believed was due to government intransigence and rampant capitalism. Sharon Beder, professor of Social Sciences at the University of Wollongong, defines the divide between activism and negotiation, citing 'light green' environmentalists as subscribing to material values and economic growth, with the environment as a resource to be treated with respect and care. They will negotiate and work within the 'dominant social paradigm'. Opposed are 'deep green' environmentalists who believe in the intrinsic worth of the environment and do not subscribe to material values; confrontation is the only way to achieve their goals.[54] While TWS and the ACF were at this time still tending towards the radical end of the spectrum, they welcomed the emergence of Bush Heritage because it could deliver for the environment using a different model.

In the 1980s conservation in Australia was aligned with left-wing politics ... but apolitical groups such as Bush Heritage and AWC offered a new way forward.

At the same time, another organisation, with much the same objectives as Bush Heritage, was being established in Western Australia. Martin Copley, a British financier and insurance underwriter, had fallen in love with Australia in 1966. In 1991 he purchased his first property, a large expanse of natural bushland at Chidlow in Western Australia, and in 1995 created Fund for Wild Australia, which later became Australian Wildlife Conservancy (AWC).[55] Copley had been inspired by South Australian mathematician John Wamsley, who had begun setting up a network of wildlife sanctuaries (Earth Sanctuaries) across Australia in 1969. Wamsley's approach was to remove the invasive weeds and eradicate all feral animals from a bushland area before erecting a fence to keep them out permanently, enabling native vegetation and native animals to flourish. He created his first earth sanctuary, Warrawong, in the Adelaide Hills, and threatened species such as the greater bilby and the burrowing bettong were soon thriving there. He built ten sanctuaries, attracting followers throughout Australia with his innovative solution, and his passion and commitment motivated thousands of people to support protections for wildlife. In 2000 he floated his enterprise on the stock exchange, but it was a financial disaster.[56] Copley bought several of Wamsley's properties, positioning AWC as the organisation to save Australia's endangered native fauna. Both AWC and Bush Heritage appealed to Earth Sanctuaries supporters so that many turned to these organisations when Earth Sanctuaries folded. However, the land conservancies suffered a credibility issue with major donors following the Earth Sanctuaries failure, and Doug Humann, when he became CEO of Bush Heritage, would have to defend the Bush Heritage model and its ambitious vision against the sceptics.

Philanthropy was slower to take hold in Australia than in the US, but by the 1990s more Australians were supporting causes or institutions in which they believed. And they were attracted to the Bush Heritage idea of land purchase as a tangible asset that would remain forever, as opposed to more abstract propositions. Signs that the environment was increasingly under pressure were becoming more obvious, as was the fear that governments were not doing enough to protect the Australian landscape. In the 1980s conservation in Australia was aligned with left-wing politics, and the divide along the lines of established wealth and capital against radical 'lefties' was clear-cut. But apolitical groups such as Bush Heritage and AWC offered a new way forward. The Franklin River protest had cemented Brown's position as an effective spokesperson, one whom people could believe and trust, although many on the right of politics saw him as too radical and a threat to economic growth and regional development. He spoke with direct simplicity and passion about what he considered important, connecting with broad-ranging audiences. While striking a chord on the need to preserve the river for future generations, he also spoke to those who believed that the natural balance of the world was changing, and that people were responsible. To them, he spoke of the 'right of the river and its environs and its ecosystems to continue to exist against a temporary destructive escapade by modern acquisitive materialist society'.[57] Speaking out, sometimes as a lone voice in the political sphere, he was significant in spelling out the need to act now.

Growing pains

By 1993 the situation at Bush Heritage was more secure. Jeremy Hobbs had set up a system to receive monthly donations so that there was now enough income to support the interest repayments on the bank loan, and the amount owing on the two blocks began to reduce. The directors began exploring possibilities for a new acquisition.

FAN PALM, 8.17 HECTARES between the mountains and the sea in the Daintree Rainforest, 'came out of the blue' and, like the Liffey blocks in Tasmania, its plight could not be ignored. A rare habitat for the threatened cassowary, the block was a crucial part of a wildlife corridor in an area that was being subdivided into 1000 household blocks. Margaret Atkinson, still shouldering much of the administrative work for Bush Heritage, was shocked while on holiday in Queensland to see development within the Daintree Forest, and by the threat to the Fan Palm block in particular.[58] The Daintree Forest Foundation had bought the block but was forced to sell it, and the Wet Tropics Management Authority (WTMA), which had bought adjacent blocks, suggested Bush Heritage buy it. Board members, keen to branch out of Tasmania, agreed, and the block was bought for $85 000 – a big commitment, since fundraising was still slow, and the Liffey blocks were not yet paid off. WTMA applied for government funding to support local community efforts to build a boardwalk through the bush, erect signs indicating new ownership and accessibility, and for appropriate management of the block. Victoria Templeton assembled another brochure with strongly emotive images highlighted by phrases such as 'Daintree Peril', 'Rainforest Rescue', 'Nothing short of spectacular', and the donation level rose dramatically.

At last Bush Heritage could justify including 'Australia' in its title. There were also salutary lessons to be learnt. Brown had foreseen problems of management for an Australia-wide organisation, given the distances and limited staff numbers. Fan Palm showed what could be done with local volunteers, while the overwhelming

FOLLOWING SPREAD **Tangled tropical forest at Fan Palm Reserve, Qld, is a haven for many species, including the nationally endangered cassowary.**

Photo: Craig Allen

response to the fundraising campaign for the Daintree block indicated that land purchase galvanised attention. But the surge in donations exposed Bush Heritage's financial vulnerability. Real business and financial skills were required to steer the organisation through the financial minefields that were beginning to surface. Robert Rosen arrived by chance at the right moment in 1993.

Rosen's career was in business and finance, but he worked in the community sector. He learnt about Bush Heritage while attempting to establish a rainforest trust in conjunction with the Nature Conservation Trust in New South Wales. When the trust failed to eventuate, he contacted Peg Putt, and 'before I knew it I had been invited to join the Board'.[59] When he became treasurer, Bush Heritage owed $188 000, had a donor database of approximately 1500, and owned three blocks of land. Although still unsure when the debt would be paid off, optimism was high, and Bush Heritage, having rediscovered the way to stimulate donations, had acquired a thirst for buying property that would not be assuaged for some time.

Rosen's inclusion on the Board was enormously beneficial, far beyond anyone's imaginings at the time. Apart from his valuable financial expertise, he had contacts in legal and accounting circles, including one with a partner in the law firm Allen Allen and Hemsley (later known as Allens Arthur Robinson (AAR), and now Allens Linklaters – hereafter referred to as 'Allens').[60]

Australian Bush Heritage Fund Board, October 1993. From left: Peg Putt, Greg Blake, Robert Rosen, Bob Brown, Penny King, Ray McKendrick (absent: Verna Simpson)

Photo: Steve Warland

Rosen had worked with John Martin, a Sydney-based partner in Allens Arthur Robinson and knew that, through him, the firm provided pro bono legal work for the NSW Nature Conservation Trust. His suggestion that AAR have a similar arrangement with Bush Heritage was greeted with enthusiasm by the partners of the firm. This partnership remains in place at the time of writing, and its value to Bush Heritage has been incalculable. Partners and staff have contributed thousands of hours of expertise, as well as personal donations, and have guided and advised on all legal matters, including establishing a legal framework that gave Bush Heritage tax deductibility in the US. Doug Humann, during his period as CEO of Bush Heritage, was reassured by the fact that Allens was watching out for Bush Heritage's interests. The close association and backing of the legal firm also added gravitas to Bush Heritage's negotiations with federal and state governments in ensuing years.

Buying land presupposed that the Board had the time and skills to assess each property on its merits, and the technical knowledge to manage it. They had been lucky so far: the Liffey blocks comprised robust, well-established forests that looked after themselves, and Fan Palm was managed by the local community. This would not always be the case, and Brown looked for someone trained in ecology. He asked Greg Blake, a young ecology student, to assist with property selection as a volunteer. By October 1993 Blake had become a Board member, and in 1995 he became Bush Heritage's first land manager.

For Blake, it was the chance of a lifetime. A self-described 'uncivilised boy from the back blocks of rural Queensland', he went to university in his mid-twenties to pursue his passion for natural history. He was 'severely dyslexic and had never handwritten more than two lines of text'. Computers and the 'indulgent support of compassionate people' such as Kirkpatrick, Brown, Simpson, Henderson and Rosen enabled his participation in cutting-edge nature conservation at a time when conventional conservation methods had stalled. Blake believed the major players were 'more interested in maintaining the aesthetic integrity of childhood walking tracks' than confronting issues of land clearance, salination and species extinction that were compromising ecological integrity in Australia. Blake focused on practical outcomes, and immersed himself in the job. Bush Heritage was the core of his life for several years, 'a source of joy and hope at a time when nature conservation was in deep crisis'.[61]

Blake's mentor, Jamie Kirkpatrick, believed that his involvement with Bush Heritage as a 'keeper of the biodiversity truth' would provide an overview of practical conservation.[62] Blake and Brown travelled the country to assess properties, and to ascertain which areas were most under threat. With limited finances, they had to analyse the environmental threats in each state to save land most at risk.

Bush Heritage, having rediscovered the way to stimulate donations, had acquired a thirst for buying property that would not be assuaged for some time.

Victoria was already well served by conservation bodies. The Bolte government's destruction of the Heytesbury forest in the 1950s for agricultural settlement, and its attempt to clear the Little Desert in the 1960s, were the last times the government authorised wholesale clearing of native forest. The RE Ross Trust, a significant philanthropic body in Victoria, wanted to support government acquisition of suitable land for parks, and Roy Everard Ross, a keen bushman and significant philanthropist, worked with the Liberal Premier, Rupert (Dick) Hamer, and the Minister for Conservation, Bill Borthwick, to establish a semi-private entity with tax deductibility for donors, to enable people to contribute to the permanent protection of land by donating either land or money for conservation. The Victorian Conservation Trust was created in 1972 to manage this work, later becoming the Trust for Nature (TfN). A similar organisation was set up in New South Wales at roughly the same time. In 1978 TfN expanded its vision to incorporate a 'revolving fund' policy, whereby TfN would buy land, covenant it and then sell it again to a sympathetic buyer, with the sale money returning to the fund to allow additional purchases. The covenant inhibited further development, and subsequent owners were required to protect the natural vegetation under covenant. TfN also encouraged farmers to consider covenanting their remnant vegetation. Its powers were further strengthened by the *Flora and Fauna Guarantee Act* of 1988, with its overarching objective of conserving Victoria's special flora and fauna. TfN has acquired and on-sold over 200 properties and helped countless other landowners covenant their land.

In Queensland clear-felling for agriculture was proceeding at the rate of 450 000 hectares a year, which meant that saving land in Queensland became the top priority. However, there would be several other purchases, and some dissension, before Bush Heritage began its journey into Queensland's remote places. For Board members the move away from the forest heartland of Tasmania was difficult.

In Queensland clear-felling for agriculture was proceeding at the rate of 450 000 hectares a year, which meant that saving land in Queensland became the top priority.

But the environmental world was changing. The word 'biodiversity' was creeping into the everyday language of conservationists and into the lexicon of the broader community, taking the idea of conservation beyond forested places, and broadening the definition of landscapes with high conservation value. Biodiversity had long been understood by scientists, botanists, zoologists and entomologists as the essential web of life, an 'intricate tapestry of interwoven life forms'.[63] It embraces the combination of millions of life forms occurring in the range of ecosystems that includes deserts, forests, wetlands, mountains, lakes, rivers, agricultural landscapes and seas. Their interactions with each other within their own ecosystems and with the rest of the environment make the earth a unique environment in which humans can live, and are 'the stuff of life' on which we totally depend.

In 1986 the National Academy of Sciences and the Smithsonian Institution held a National Forum in Washington DC, entitled 'Biodiversity'. Over sixty leading biologists, economists, agricultural experts, philosophers and other professionals participated, with an audience of several hundred people, garnering wide press coverage. Interest in biodiversity-related issues had been growing since 1980, as data from scientists in the different disciplines of deforestation, species extinction and tropical biology accumulated. This effectively highlighted global problems, where connections between the conservation of biodiversity and economic development were impossible to ignore.[64] David Ehrenfeld, professor of Biology at Rutgers University, described this change in attitude:

> Two to three decades ago, the topic [biodiversity] would not have been thought worth discussing, because few scientists and fewer laymen believed that biological diversity was – or could be – endangered in its totality. Three or four decades before that, a discussion of the value of biological diversity would probably have been scorned for a different reason. In the early part of the twentieth century, that value would have been taken for granted: the diversity of life was considered an integral part of life, and one of the nicest parts at that. Valuing diversity would, I suspect, have been thought both presumptuous and a terrible waste of time.[65]

In 1988, recognition that ecosystems and individual species were increasingly threatened because of human activity came to a head in the United Nations (UN) Environment Programme with the Ad Hoc Working Group of Experts on Biological Diversity. This paved the way, through many meetings and committees, for the UN Conference on Environment and Development (the Rio 'Earth Summit') in 1992.

The Rio Earth Summit was an important step.[66] For the first time sustainable economic development was officially and internationally linked to the preservation of ecosystems, species and genetic resources. Australia was actively involved in developing the Convention on Biological Diversity, and was an early signatory to the Rio convention. It committed to Agenda 21, the international framework agreement for pursuing global sustainable development, established to ensure countries met their obligations to protect the world's biodiversity. Agenda 21 identified biodiversity hotspots around the world under threat, and Australia scored one, in south-western Western Australia, a region of unique and biodiverse richness.

In his Environment Statement of December 1992, Prime Minister Keating declared that:

> the establishment of a comprehensive system of protected areas is vital if we are to retain our status as a region of megadiversity. Besides being ecologically viable, these areas must represent the full range of ecosystems. The Government is committed to the development of a national comprehensive system of parks and reserves. This will be achieved in cooperation with States and Territories.[67]

That cooperation led to the establishment of the National Reserve System (NRS), developed from the rolling out of the Interim Biogeographic Regionalisation for Australia, which would be updated as spatial mapping and information on vegetation communities and ecosystems from the state and territory agencies became available. Initially the NRS offered one dollar for every one dollar raised to buy private land that fitted the prescribed biodiversity criteria. This funding came as a godsend to environmental groups, and from 1995 Bush Heritage began to look for larger properties, a move that would

Janice Bird and Greg Blake in the office

Source: Bush Heritage Australia

change its vision and impact significantly on conservation in Australia. In 1999, Tarcutta Hills Reserve in New South Wales was the first property purchased by Bush Heritage using NRS funding.

Bush Heritage's timing was perfect. It was incorporated just as Labor prime ministers Hawke and Keating were withdrawing from adding more land to the Commonwealth parks estate, choosing instead to establish the NRS. The program encouraged and supported other organisations to buy the land that fulfilled Rio Convention requirements, leaving those organisations with the long-term management of the properties. Between 1992 and 1996, the federal government spent $11.5 million on the National Reserve System Cooperative Program, and from 1996 accelerated the drive to buy land by investing millions of dollars to help the states, territories, Aboriginal communities, local councils and conservation organisations to buy and establish conservation on lands for the reserve system. Of the state governments, the SA government supported private purchase of threatened land, thus avoiding the prohibitive costs involved in continued maintenance, as will be seen with later Bush Heritage purchases in South Australia.

Australian Bush Heritage Fund Board, 1995. From left: Ray McKendrick (back), Robert Rosen, Greg Blake, Bob Brown, Judy Henderson, Verna Simpson, Peg Putt, Penny Figgis, Penny King

Source: Bush Heritage Australia

The other states were not enthusiastic about NRS money going to private entities, believing that these organisations would not last and thus would not be able to manage land for the long term.

By the time the NRS hit its stride, Bush Heritage had become a major player buying privately owned pastoral leases. Some properties were degraded and economically unviable, but nevertheless retained elements worth conserving. By 2011, Bush Heritage had received $8.137 million of NRS funding and had purchased eleven properties, protecting 848 000 hectares of land. It is not surprising therefore that Bush Heritage – with WWF, TNC, TfN, GA, Tasmanian Land Conservancy (TLC), ACF and the Pew Charitable Trusts – continued to lobby the federal government to provide ongoing support for the NRS. But the funding dried up in 2011 under Labor's Minister for Conservation, Tony Burke, who diverted funds to buy back Coral Sea fishing licences, and for settlement with Tasmanian Forestry over Tasmanian forests. With no allocation of money for general conservation purposes, it was easy in 2013 for Prime Minister Tony Abbott's Coalition government to deny the NRS further support. But this was in the future; in 1993, Bush Heritage still had its heart in Tasmania's forests.

In 1993 Bob Brown was still Bush Heritage's significant leader, although this was about to change. In the early years those who worked for Bush Heritage were inspired by Brown and his belief that they could achieve something important for the environment, and it was this era that set the tone for the organisation. Everyone knew they were the 'new kids on the block' in the conservation world, and had something to prove. Because they were so few, they worked together, supported each other and cooperated, alive to and tolerant of each other's personal idiosyncrasies in the way families accept differences. That sense of family still pervades Bush Heritage today, despite its size and the fact that, by necessity, there is a more corporate ambience. The inclusivity that defined Brown's leadership is still influential, and has been reinforced by successive presidents and CEOs in their own ways. Like many NGOs Bush Heritage seems to attract people who sacrifice high financial reward and high-profile career opportunities, and are prepared to work for positive environmental outcomes in a collegial and sympathetic ambience.

Change came as Bush Heritage settled into its own skin, and began to flex muscle in the national arena of conservation. Like every organisation that survives, Bush Heritage went through different stages as the vision was fleshed out and shaped by outside forces. The period between 1993 and 1999 was one of growth and consolidation, positioning Bush Heritage to take the next big leap forward into the twenty-first century.

That sense of family still pervades Bush Heritage today, despite its size ... The inclusivity that defined Brown's leadership is still influential.

Victoria Templeton's departure meant changes in the day-to-day workings of the office. As a stop-gap measure, Judy Henderson resigned from the Board to manage the office. During this time, Bush Heritage achieved a level of calm and financial stability, with directors keen to expand the Board's reach and to employ more people to handle the increasing workloads. But Henderson was juggling other commitments, so in 1994 she and Brown employed Janice Bird as national coordinator, allowing Henderson to retire from the everyday running of the organisation.[68]

Bird's involvement with nature conservation began in England when, as a volunteer, she set up a local branch of WWF. On visiting Australia she found inspiration and joy in Australia's wild places – Uluru and Kakadu – in the Northern Territory. She joined the Franklin River campaign in 1982, moving permanently to Tasmania in 1993, where she worked at TWS, before transferring her allegiance to Bush Heritage. Like Templeton, she found the work exhilarating and revelled in the variety of tasks: liaising with Blake for property acquisitions; organising rate exemption applications with councils; handling changes in land tenure, conservation agreements with state departments; applications for reserves to become wildlife refuges; and all the bureaucratic correspondence Bush Heritage was learning went with ownership of conservation properties. While Atkinson continued to manage the donations, Bird dealt with the rest of the administration, including interviews with the media and presentations to companies and trusts in Sydney and Melbourne. She enjoyed an easy rapport with donors, and promoted the bequest program, which would prove valuable for Bush Heritage in the future.[69]

Donor generosity impressed her, too; she remembers holding a cheque for $30 000, knowing she had never held so much money before.

Less productive was the Community Service Announcements strategy. All the free-to-air television networks agreed to run a segment featuring designer Jenny Kee, filmed by Bush Heritage in bushland near her Blue Mountains home. Unfortunately, it was shown late at night, tucked in among advertisements for phone sex. As those advertisements had a premium-rate phone number and Bush Heritage a freecall 1800 number, viewers often offloaded their sexual frustrations on the Bush Heritage answering machine. 'I always knew when the ad had been shown because next morning I'd be greeted with a frantically flashing light on the answering machine. I listened, in case there was a genuine supporter's call, but it was a horrible way to start my day.'[70]

Bird, shouldering a huge workload, was smart enough to ask for more help. With the arrival of another staff member in the office, larger premises were needed. They moved to a second-storey office on the corner of Murray and Collins streets, still in the heart of Hobart, a space they initially shared with the Environmental Defenders Office. They soon had to enlarge these premises by knocking through the wall for access to a larger office with big windows overlooking the street. It was liberating until Bird realised how much she gesticulated while on the phone as she stood looking out the window. She was disconcerted when people at the busy intersection below waved to her, looking puzzled. 'In Hobart, you expected to recognise everyone you saw on the street, so passers-by who saw me waving assumed they must know me and waved back.' It was an interesting comment about Tasmania's capital city, the size of which would influence Bush Heritage's decision to leave Hobart for the mainland later on.[71]

Margaret Graham, the new administrative assistant, was also energised by the palpable enthusiam of the Bush Heritage office. The generosity of donors impressed her, too; she remembers holding a cheque for $30 000, knowing she had never held so much money before. 'It was a fabulous place to work, really collegial, with no sense of directors and employees. We were all in it together.'[72]

As computers became more important for office administration during the 1990s, Bush Heritage needed to upgrade its database of supporters; it also needed more employees and more Board members with particular skills. Judy Henderson had stepped back, and Jeremy Hobbs, Penny King and Verna Simpson had departed or were reducing their involvement, paving the way for newcomers. And in 1996 Bob Brown stood down as President, 'after three booming years'.[73] His political life was taking up more of his time, and it was felt that his politics might have a counterproductive effect on potential donors, generally believed to be conservative, although this was an overgeneralisation! Brown would remain on the Board, but it was time to appoint a new president, with new ideas.

Judy Henderson already knew who that person might be. Karen Alexander had campaigned on the Franklin River, her career was in conservation, and she was a Bush Heritage donor. When in Tasmania, she usually stayed with Henderson, and was always keen to hear how the organisation was faring. She responded eagerly to Henderson's suggestion that she lead a strategic planning session with the Board to develop a plan for the future direction of Bush Heritage.

Establishing guidelines for bush purchases

Following the success of her planning session with the Board on possible future directions, Karen Alexander was invited to fill a casual vacancy on the Board and on the resignation of Bob Brown in 1996, was subsequently elected the new President.

THE BOARD WAS COMPOSED of people who might have seen themselves as left-wing greenies, and this did not alter with Karen Alexander's appointment; her networks within the environment movement were impressive, and supported her move. In truth, the conservation movement was one big cross-referencing of old friends and foes, who employed each other, recommended each other for jobs, and campaigned together, collecting many different job titles over the years. When Alexander began to consider people to join the Board, she would look to these friends and colleagues for the next generation of Board members.

By the time Alexander became President she had already experienced most aspects of the environmental movement. A veteran activist, 'engaging, charismatic, vivacious and one of the clearest voices of reason', she had spent four years full time as a volunteer on the Franklin River campaign, helped TWS extend its reach beyond Tasmania, and assisted in establishing the Greens political party in Victoria.[74] She also founded the Orienteering Association in Tasmania, and worked in France on the UN Environment Programme, followed by three years as Campaign Director for ACF. Her talent in bringing people with divergent views together was legendary, and it was her strategic vision, energy and ability to get on with everyone that helped drive Bush Heritage to the next level. Her leadership style was subtle. An excellent chair of meetings, she preferred a background role to the limelight, listening to directors talk through issues, summarising and adding anything she thought was missing, asking extra questions. It was not until Alexander's last meeting as President that colleague Mara Bún realised how skilled she was in asking the questions that no one else did, as if by accident![75] For Bush Heritage, Alexander offered new energy that did not stray far from the familiar.

Bush Heritage Board, September 1997. From left: Lincoln Siliakus, Margaret Robertson, Penny Figgis, Karen Alexander (President), Ray McKendrick, Robert Rosen, Verna Simpson, Judy Henderson (absent: Mara Bún, Louise Gilfedder)

Source: Bush Heritage Australia

When Bush Heritage paid off the Liffey blocks in 1995, the Board was ready to make a new purchase, with land manager Greg Blake leading the process. Since Bush Heritage was the first private organisation to concentrate solely on acquisition and management of land across Australia, Blake wanted to develop an effective model for others to follow. While he and Bob Brown spent time exploring the potential of land throughout Australia, they were inhibited for some time from actually buying another property, because they believed it was important to establish a set of criteria on which to base any new purchase.

Blake took advice from those with more experience in ecology and the environment, one of whom was Paul Sattler. A former senior bureaucrat in the Queensland Department of Treasury and Finance, Sattler had 'got nature conservation the way some people get religion'.[76] He transferred to the Queensland National Parks Service, and from 1976 managed the Department of Environment and Heritage's acquisition program. While Blake was gathering material to establish guidelines for property purchase, Sattler was advocating a bioregional approach, coupled with a systematic assessment of conservation data, so that all bioregions could be considered simultaneously. The 1999 book, *The Conservation Status of Queensland's Bioregional Ecosystems*, co-edited by Sattler and Rebecca Williams, refuted the Queensland government's claims that it had effective conservation controls in place. The preprint review of the book caused such controversy that Sattler was forbidden by the government to publish it. He ignored the ban, and the book helped instigate the discussion that brought about change in Queensland's land clearance legislation, scaling back land clearing from the previously accepted rate of 450 000 hectares of bush each year – and it was significant environmental legislation at the time.

Blake also drew on current leading scientific information to develop a process that Bush Heritage used, with modifications, for some years, although it would be some time before it became the only basis for buying land. Early on, purchases were more likely to happen because of the personal interest of Board members, and management proposals were retrofitted to suit the property, rather than established at the outset.

The guidelines for acquiring properties that Blake and his team developed over several years use a process modelled on the scientific framework known as CAR, established by the NRS program to protect ecosystems. CAR, an acronym for 'comprehensive, adequate and representative', was endorsed by all Australian governments. 'Comprehensive' meant 'the inclusion of examples of regional-scale ecosystems in each bioregion'; 'Adequate' meant 'the inclusion of sufficient levels of each ecosystem within the protected area network to provide ecological viability and to maintain the integrity of populations, species and communities'; and 'Representative' called for the 'inclusion of areas at a finer scale, to encompass the variability of habitat within ecosystems'.[77]

Ideally, a property search would begin with reference to academics, local interest groups, and the local park services that maintained a wish list of properties they would like to buy. Alternatively, someone in the organisation might know someone who wanted to sell a property, as in the case of Fan Palm, and there was always a steady stream of letters from farmers or property owners who wanted to sell. They were often farmers who had struggled for years without success, were going bankrupt, or had no descendants who wanted to continue working on the farm. There were also those who genuinely wanted to see conservation initiatives established on their land. Blake and Jos Bentley, who joined Bush Heritage as an ecologist in 1998, also developed regular contacts with rural property agents, since properties often changed hands quickly. Having found a likely property, Bush Heritage had to consider its national, regional and local significance before ascertaining its viability.

The first thing was to assess the threat to the area, since it was pointless to use scarce resources to protect something not at risk. The most threatened ecosystems were usually on lands used for cropping and grazing, since national parks were established on land with little or no commercial value. The threats would then be quantified and qualified, taking into consideration world markets for commodities, new technologies that might be introduced to exploit virgin soil types and landscapes, new fertilisers, genetically modified crops, and current and future threats that could come from river systems both downstream and upstream. Once the threat was identified as sufficiently serious, they would assess the viability of trying to restore the property, since if the degradation was too severe, it might not be worth buying. Assessing conservation values and threats was time-consuming, specialised work, and Blake maintained academic connections with ecologists, botanists and zoologists, soil and climate experts who provided their help, often gratis, in return for opportunities to visit out-of-the-way properties in remote Australia. Having satisfied these criteria, the property would then be rated against other purchase options to ascertain best value for money. If possible, neighbours' reactions to the idea of a conservation organisation next door were also gauged. Ecological reports on a property would be collated and synthesised into a single report and presented to the Board for consideration. The first time this assessment process was used effectively was for the purchase of Carnarvon Station, and that would not be until 2001.

The property also had to pass the financial test. Could they afford it? How easy would it be to raise funds?

And the purchase price was not the only issue to consider. The money to be raised had to include management costs for the first three years: staffing, fencing, weed eradication, fire management, road access and administration.[78] Assessments then had to be made on all possible issues, and costs assigned. But this might not be enough to get it over the line. It was possible that a wonderful property, too good to pass up, might become available and knock the property already under consideration out of contention. This last became increasingly unlikely after the management team integrated the data on properties for sale into the Bush Heritage system. The data were comprehensive, and the team reviewed this information every month.

Once a decision was made, negotiations could begin. Blake knew these dealings could be protracted, without assurance of a positive outcome. Board members had to hold to the commercial value of the property in order to avoid overpaying and thus creating a high 'conservation' price. But Blake and Brown also believed they ought to explain to potential vendors that properties with high nature conservation significance could have a financial value in themselves: that it paid to maintain the natural vegetation, and to protect the land from erosion and other degradation; that clearing the land was the least positive option. This could result in vendors reconsidering their options and withdrawing their properties from sale.

At the same time, the fundraisers had to prepare for a fundraising campaign to secure funds for the purchase and for ongoing management. Again, Blake foresaw problems if they could not purchase because funds were insufficient or, if funds were available but the property deal failed, the close connection that might have been assiduously developed with donors over a particular piece of land could be compromised. But, as he later commented, this turned out to be an irrational fear, since Bush Heritage donors were 'the least emotionally random people with which any conservation organisation had to deal'. The organisation seemed to attract 'smart, pragmatic' people who appreciated honest explanations that acknowledged the risks, without the spin.[79] As the broader issues of conservation were spelled out in quarterly newsletters, the support base grew stronger and more focused.

Going national

In 1995 Greg Blake took a phone call from an elderly man that put the focus squarely back on buying more land. He had recently come across a Bush Heritage brochure about saving Fan Palm, and said that his wife loved koalas and was 'hassling' him to give money for their survival.

THEY CHATTED ABOUT MANY THINGS: sailing boats, Sydney, politics. Several days later a letter arrived, from which two cheques totalling $200 000 fell out. The first cheque, Blake remembers 'literally made my legs go weak and my head spin', and with the second cheque, 'I nearly fainted'.[80] Janice Bird danced with joy, recognising 'a real turning point'.[81] It was the first six-figure donation that the organisation had received. In 1995, such a donation potentially covered the cost of a new property; it would allow Bush Heritage to achieve a credible position on the mainland without having to dig too deep financially. Both Blake and Bird believe that, without that donation, Bush Heritage might not exist today. The NSW couple continued to generously support Bush Heritage for many years.

The hunt was on for koala habitat. Blake and local NSW ecologist Jackie Miles, using advice from the National Parks and Wildlife Service and other locals, found Brogo, 120 hectares deep in the valley of the Brogo River, 20 kilometres north of Bega on the NSW south coast. The Board agreed to the purchase. Prior to this Blake took Bob Brown and Robert Rosen to view Brogo, and, as they walked the boundaries, a tough trek through thick forest, up steep hills and down to valley floors, the beauty of the landscape confirmed for them its value. Part of a subsidiary range that connects to the Great Divide through a series of ravines and rocky cliffs, Brogo marks the northern point for several species of birds migrating from the south and the southern point for some northern migrants.[82] The previous owners, a retired Dutch couple who bred angora goats, remained good neighbours after selling half their property to Bush Heritage. Blake's research revealed that the reserve was one of the largest areas of intact native bushland in the region, with several forest ecosystems that were poorly protected elsewhere. Ninety per cent of dry rainforests in Australia had already been lost, and at Brogo the dry rainforest includes native fig trees that twist into fissures in the outcrops of granite boulders, shading the broad-leafed orchids that flourish on top of the rocks, and put forth a profusion of long, yellow-flowering stems in spring. The fig trees nourish fruit-eating birds and bats that spread the seeds far and wide. Other important ecosystems, the riverine forest and scrub near the river, comprise river sheoaks, gossamer wattle, tree violet and basket grass, and wet vine forests. They support over seventy species of birds, the sooty, powerful and boobook owls gravitating to the older trees where hollows provide habitat for the abundant birdlife and other creatures.

Although the Brogo property was not as big as the Liffey Valley reserves, it meant another step onto the mainland, encouraging the search for more properties. Blake travelled around Australia, assessing possibilities, particularly in Queensland. Pastoralists there were clearing land at an even greater rate, trying to beat the legislation limiting clearance that they knew was coming.

But events in Western Australia diverted attention from Queensland. Margaret Robertson, now a Board member, received a letter originally sent to the WA Conservation Council by Kojonup farmers Phillip and Anthony O'Halloran.[83] They had tried for ten years to sell their 333 hectare block of native bushland to the state government through the Department of Conservation and Land Management (CALM), but, despite being interested, the Council could not afford it. Beth Schultz, from the Council, knew Robertson's connection to Bush Heritage and passed the letter to her. By coincidence, the family who wanted to sell were neighbours of Robertson's parents, who were still farming near Kojonup.[84]

Kojonup on the Albany Highway is the centre of a farming community, where land clearing continued until

Brogo River Reserve delivered much more than koala habitat.
Photo: Wayne Lawler/Ecopix

Thousands of hectares of land were rolled and burnt with no thought of the long-term consequences.

Photo: Geoff Bee

the 1960s. Wheat farms stretch forever, on land that was first used for cattle and sheep properties. In September 2002, the Kodja Place was opened. A community project that brought people together, it was established to tell the story of settlement through the words and photographs of the people who were there: the Noongar people, the white settlers, and the European immigrants who came straight from the ships seeking work, a poignant reminder of the effort that went into creating agricultural pasture land. The interpretive display about creating farms is especially relevant to the issue of bushland being destroyed for a greater good, and in one of the panels, William Harvey, from Orchid Valley, describes the mood of the times:

> People took out bank loans and bulldozers hit the area so you started to get big quantities of land knocked down. When we were first married in the early '70s we knocked 6000–7000 acres of this heavy timbered country down in a year. There were thousands of tons of timber turned into smoke. We thought nothing of it; we were carving out a future in agriculture. At the time, no one was aware of the implications of what we were doing, we got a tax deduction for it – it was encouraged. You know, the whole future of the nation depended on us doing this.[85]

Wandoo gum woodland, characterised by tall, thick-trunked trees with white bark and a varied understorey of open shrubland, once covered the landscape. When clearing ceased, little bush was left in the Kojonup area, so the 333 hectares that the O'Halloran brothers wanted to save was special.[86]

The woodland had never been grazed, because the poison pea on the site was fatal to stock. The brothers did not want to sell to a farmer who would clear it for pastureland, and Robertson saw an opportunity for Bush Heritage. Blake assessed the property, and Robertson canvassed interest from the local community, who proved keen to rescue a small piece of virgin bushland. The process to buy the property began, driven by community goodwill and energy.[87]

Blake, however, was not enthusiastic: one corner of the property had a serious salinity problem that would spread, and he spoke against the purchase at the Board meeting.[88]

Kojonup's pride in protecting its precious remnant bushland is evident in the roadside sign.

Source: Doug Humann

But Robertson, with the Kojonup community supporting, pushed hard, and the Kojonup purchase went ahead. Robertson knows that in 2014 Bush Heritage would not have bought the site: it was too small and the conservation values would not meet current criteria. But the deed was done. O'Halloran's letter arrived in 1995, settlement occurred in 1996, and the purchase was celebrated with a picnic on the property and an address by Bob Brown in March 1997, attended by local community members, including farmers and naturalists, representatives from the Shire, CALM staff and representatives of Agriculture WA, who all wanted to be involved in the management of the block. The Bush Heritage newsletter headline – 'WA, Here We Come!' – heralded a new beginning. And while Bush Heritage was introduced to conservation in Western Australia, WA conservationists had also discovered Bush Heritage; possibilities for further cooperation were exciting. The Kojonup community expressed its support by erecting a sign on the road: 'Welcome to Shire of Kojunup: First shire with a Bush Heritage reserve'.

The Kojonup purchase was important for several reasons. Once again it demonstrated that Bush Heritage was serious about being an Australia-wide organisation, and that distant properties could be managed by informal partnerships, with local community support. More importantly, it sparked efforts by an enthusiastic group formed from the WA government departments of Agriculture and Sustainability, to find properties that Bush Heritage might buy.

Keith Bradby was one of the group's most outspoken and innovative members. He ran a small company, Bush Brokers, that brought farmers and buyers together, and he factored in Bush Heritage as a possible buyer for farmland property to be saved for conservation.[89] The Bush Heritage Western Australian Support Group met several times in 1997 to discuss Bush Heritage involvement, although discussions took several years to mature. In the meantime, they brought Margaret Robertson and Keith Bradby together and cemented the relationship that led to their marriage in 2000.

Kojonup Reserve, WA, a habitat island in a sea of cleared land

Photo: Simon Neville

A CEO for Bush Heritage

In 1997 Bush Heritage had just finalised the $20 000 purchase of the grazing lease for the wild and beautiful Erith Island in the middle of Bass Strait. 'An astonishingly harsh and beautiful place, full of screaming gales and haunting silences', historically rich in Aboriginal culture, used extensively for fur sealing, and a graveyard for ships falling foul of its rocky coastline, Erith Island's history is rich and sometimes dark.[90]

SINCE 1962 THE ISLAND HAD BEEN the holiday destination of the Murray-Smith family. Stephen, a teacher, historian and founder-editor of Melbourne literary journal *Overland*, and Nita his wife, spent summers on the island with family and friends, revelling in the isolation and closeness to nature. Artist Fred Williams was a friend, and among several paintings he did at the time, his fluid and evocative *Beachscape I* and *II* capture the magic of the place. The 'Erith Island mob' maintained the hut, and cut minimal tracks across the island. After Murray-Smith's death, the family wanted to ensure that the island remained unspoilt. Chris Robinson, a conservationist, Greenpeace activist, and one of the original *Rainbow Warrior* crew, persuaded Greg Blake that Bush Heritage must help in saving the island.

The mystique of Erith Island raised Bush Heritage's profile and added to its financial reserves. For Nita Murray-Smith, the Bush Heritage purchase of her beloved island was 'more than we could have hoped for'.[91] For Blake and friends, looking after Erith Island was a wonderful adventure: sailing across the strait with Robinson on his yacht *Fand* to eradicate feral animals and weeds, and to party with a growing number of volunteers. They documented rare plants and listed sightings of herons, flame robins and and grey fantails, among other birds; they revelled in the proximity of dolphins in the bay, while *Fand* rode high in the narrow cove, where towering granite cliffs clothed in rusty orange lichen sheltered it from tumultuous ocean swells. James Button, an *Age* reporter who spent time on the island with Blake and a group of volunteers, found it a 'wild, unnerving, utterly peaceful place'.[92]

Because of its remoteness and the dangerous crossing, owning the lease was a real challenge for Bush Heritage. On top of this was the additional burden of Deal Island. As the leaseholder of Erith Island, it was difficult for Bush

Heritage not to agree, in late 1998, to also take over the lease of nearby Deal Island, on behalf of the Tasmanian Parks and Wildlife Services. That lease continued until 2000, before being transferred back to the government. In 2002 the Erith Island lease was sold back to the government to facilitate the declaration of the Kent Group National Park. In 2014 Bob Brown acknowledged that without Bush Heritage's ownership and work in clearing the islands of feral animals and weeds, this outcome might never have been achieved.[93] Today, Erith Island and Deal Island are protected forever.

At the same time, Bush Heritage was expanding to cope with the new properties and those listed for possible acquisition. To Karen Alexander, flying from Melbourne to chair Board meetings and sort out staff problems, it was apparent that the Bush Heritage model was not working efficiently. The Board was still too involved in running the organisation: actions were not coordinated and there was no clear direction. The Board needed to distance itself from day-to-day activities. Alexander envisaged an executive director to whom staff would report, and who would bring projects and staff concerns to the Board. The position was advertised in May 1997, and Alexander had a candidate in mind. She contacted Doug Humann and suggested he consider the role.[94]

At the time, Humann, Executive Director of the VNPA since 1990, was dealing with Premier Jeff Kennett's plans to open up parks for development, and was feeling the frustration of having to fight for the continuing existence of land already designated as parks or nature reserves. When Alexander suggested he join Bush Heritage, he was keen to find out more.[95] He already knew something about Tasmania's wildernesses. His brother had been lost for five days on the Cracroft River in the rugged south-west in 1972, the same year that his father donated money towards production of a film about Lake Pedder, so he understood the pull of the wild Tasmanian landscape. But his early boyhood interest in conservation came from an old friend, Dr Alex Scholes, a teacher and dairy farmer in the upper Yarra Valley near Melbourne. As a teenager, Humann earned pocket money working for Scholes, and was influenced by his vision of a corridor in the Yarra Valley that could join farming land and state forests to the alpine region, linking conservation-focused land management with other land uses and highlighting the importance of collaborative, rather than oppositional, relationships.[96] After university, Humann travelled through Central, northern and Western Australia, visiting many remote Aboriginal communities and learning something of the relationship Aboriginal people had with the land. That experience, and a long-term friendship with a woman from the Stolen Generation, awakened a desire to work in conservation in ways that acknowledged Aboriginal rights and Aboriginal expertise in managing country.

At the same time, Bush Heritage was expanding to cope with the new properties and those listed for possible acquisition.

In 1990, believing there needed to be more focus on national parks, Humann left his job as a geography teacher and became the first full-time Executive Director of the VNPA, lobbying for more parks and better park management, but also encouraging VNPA to look beyond the parks model. By the early 1990s, funding to public protected areas was declining and fewer parks were being extended. Innovations were clearly needed. Humann's thoughts had turned more and more to the private land estate and the role it could play in helping conserve biodiversity across Australia. In 1997, Bush Heritage seemed to be offering him the opportunity to more fully explore this possibility. He and his wife Jill visited Hobart, and although all their family and friends were on the mainland, decided to give Bush Heritage and Tasmania a go. It marked the beginning of a new era for Bush Heritage. Humann took up the position in August.

Being Executive Director required tenacity and self-confidence; he had to soothe Bush Heritage's growing pains, and steer a purposeful way forward. In 1996 Bush Heritage income through donations was $30 000 and in

Erith Island, an iconic but challenging property to manage

Photo: David Neilson

1997 it was $79 000; land management costs rose from $15 393 in 1996 to $44 260 in 1997.[97] The need for smarter, more focused fundraising, new strategies and different expertise was becoming obvious, and Board members wanted to see a more disciplined approach from staff. Although some intervention was essential, there was a tendency for the Board to ignore the fact that the professionalism of its employees was compromised by heavy workloads that necessitated cutting corners to achieve the desired outcomes. Humann was the buffer for a Board gradually shifting from a hands-on approach to one of setting organisational strategy, and this was not easy. The 1990s had ushered in changes in business structures and administrative procedures throughout the Western world, and companies struggled to deal with new technology and more sophisticated business practices. Computers and the internet were revolutionising work habits, and the days when an organisation could 'muddle through' were disappearing. The bottom line, accountability, cost effectiveness, responsiveness to one's constituency, and human resource management were issues with which a fledgling business had to come to terms before it could fly, and Bush Heritage was no different from other companies as it sought to master new systems. As well, it had always relied on staff finding their reward in their work rather than in their remuneration. Boardroom tensions about directions were matched by office tensions, mostly about the long work hours. Humann, arriving from a more structured work background where staff consulted with their leaders and took direction, bumped up against an experienced land manager who was constantly on the move, used to

Doug Humann (centre) on Deal Island in 1998 with Tasmanian Premier Jim Bacon (right)

Photo: Davies Brothers Ltd

autonomy and to making decisions without recourse to a higher authority. Humann's arrival in 1997 also coincided with Board discussions about the lease on Erith Island, which proved a divisive issue.

Erith Island was always a difficult site to manage, requiring a buccaneering style that could only be enjoyed by those with access to yachts, and with a stomach for stormy seas and 4 metre waves crashing over the deck. The crossing could be dangerous, and one employee, Nathan Males, was washed overboard and nearly drowned, before being rescued by footballers on a luxury boat in the middle of Bass Strait. The incident rang alarm bells for Humann about health and safety issues that had not been addressed.

The Deal Island lease created new worries for Humann. They had to charter light planes to make the crossing, and the lease required Bush Heritage to maintain the airstrip for aircraft traffic. Humann had sleepless nights worrying about the safety of Bush Heritage employees, and what amount of care constituted a safe airstrip. Discussions with lawyers at Allens confirmed his concerns. The first safety regulations were drawn up, although Humann acknowledges they were largely reactive. It would take more property purchases and more reserve managers living in remote places for development of a comprehensive plan to safeguard its employees and assets throughout Australia.

Later in 1997, the acquisition of Friendly Beaches Reserve in eastern Tasmania, 250 hectares of coastal bush along a spectacular wild shoreline, gave Humann the opportunity to launch a new property. The purchase had its origins in Bob Brown's political work in the Tasmanian parliament, where, in 1989, he was leader of the five newly elected Greens in the House of Assembly. Besides doubling the size of the Tasmanian Wilderness World Heritage area, the Greens had also negotiated the establishment of the 16 000 hectare Douglas Apsley National Park on the east coast.[98] However, because of a sand mining proposal, the state Labor government would not extend Freycinet National Park north to include the area known as Friendly Beaches. When the Accord between Labor and the Greens broke down and an election was called, Brown persuaded the then Minister for National Parks, Harry Holgate, to sign the proclamation to enable an extension of the park to cover the northern area. In later years it was Holgate's pleasure to recall saving the Beaches, counting it the best thing he ever did.[99] However, the proclamation did not cover two

FOLLOWING SPREAD Friendly Beaches Reserve, Tas, protects dunes, coastal heathland and forest.

Photo: Wayne Lawler/Ecopix

enclaves in the centre. One property was purchased to create an eco-resort, and Bush Heritage bought the other, completing the protection of Friendly Beaches. It was a popular purchase, with Tasmanian donors keen to support a much-loved area.

Humann's appointment lifted Bush Heritage's profile, particularly in Victoria where he had been a respected leader of VNPA. Being voted 'Environmentalist of the Year' in 1997 by *Wild Magazine* helped too. People were intrigued to find out why he had resigned from a highly regarded position to work for a little-known organisation in Tasmania. On investigation, some liked what they read about Bush Heritage and transferred their allegiance, or added Bush Heritage to their list of worthy causes.[100] Humann continued to build his network of friends, business contacts, government officials and political leaders, and their value to Bush Heritage cannot be overestimated. His interest in Aboriginal people meant that from now on Bush Heritage's conservation goals would include the idea of working more closely in partnership with different clans as the opportunities arose, although it would be some time before this approach manifested in any practical outcomes.

Humann's appointment lifted Bush Heritage's profile, particularly in Victoria where he had been a respected leader of VNPA.

Working with VNPA had given Humann useful insights into government procedures. He had participated in the early work setting up the NRS program, established with bipartisan support towards the end of Keating's time as prime minister. John Howard, on becoming prime minister in a Liberal–Coalition government, authorised the sale of Telstra in two tranches in 1997 and 1999, committing the proceeds to a fund, the National Heritage Trust (NHT), for environment protection, and appointed Robert Hill as Environment Minister. This was an inspired choice, and, as it turned out, perhaps the greatest windfall that the environment movement would ever experience. Hill, who knew nothing about the environment, took advice from experts, among them the Wentworth Group of Concerned Scientists, and from organisations already involved in working for the environment. He soon formed the opinon that supporting conservation of landscapes was the 'courageous big political right thing to do' despite being accused by a senior NSW public servant of 'turning the NRS program into a marginal seat slush fund to build local council playing fields'.[101] The NRS had begun by paying one dollar for every dollar raised privately for land acquisition, but when convinced by experienced NGOs that the scheme would fail because donors were not prepared to pay for management costs, Hill responded by raising the pay ratio to two dollars for every dollar to cover management. The NRS thus became the vehicle by which the funds from NHT were disbursed for environmental purchases. Humann had worked with Howard and David Kemp, Minister for the Environment and Heritage, and he maintained good relationships with the federal environment ministers who followed, Robert Hill and Ian Campbell. The Bush Heritage approach fitted the Liberal ideology of the time, and the fact that NRS funds could leverage private investment provided enormous impetus. Humann was always surprised at how favourably applications for grants were received, and Bush Heritage benefited greatly from the NRS, particularly when the grant became two dollars for every philanthropic dollar privately raised.[102]

Soon after Humann's arrival, several new staff were appointed. Nathan Males was engaged as property manager to develop Bush Heritage's capacity in this area.[103] Janice Bird acquired an assistant – administrative officer Kerrie Green, and ecologists Philip Cullen and Jim Mulcahy were also employed.

As members completed their terms on the Board, Alexander was able to suggest new members with particular expertise. Louise Gilfedder, a botanist specialising in grasslands, was a science-focused conservation strategist involved with cultural management of natural systems. Mara Bún was manager of Policy and Public Affairs at the Australian Consumers Association, with a strong background in marketing and corporate affairs in the non-governmental sector, and Penelope Figgis, a political scientist who had joined the Board in 1995, was a national lobbyist for the ACF in the 1980s, and a powerful advocate of Protected Areas policy, World Heritage and conservation on private land. Following a further review of the skills required on the Board, Clem Doherty, a retired director of McKinsey & Company, where he led the Asia-Pacific telecommunications, electronics and multimedia division, joined the Board in 1996, as did Libby Smith, a professional market researcher, and Janice Carpenter, the owner and director of an ethical investment company.

In 1997 Bush Heritage turned its attention to Queensland. Blake contacted Rod Fensham with whom he had studied at the University of Tasmania. Fensham was working at Queensland University as a quantitative field ecologist ranging across a variety of landscapes. He also advised the Queensland government on developing policy to minimise the extinction of plant and animal species and to better manage its natural resources. He would find his niche at the Queensland Herbarium, but in the early years his expertise in identifying and mapping areas of conservation significance helped build Queensland's legislative framework to protect the natural vegetation. He and Blake worked together on these projects.[104] This legislation would not be in place until 2003, and meanwhile saving natural bushland before it was destroyed was an urgent task. Fensham's familiarity with pastoral properties across Queensland convinced the Bush Heritage Board that Queensland properties were both vulnerable and cheap.

Fensham discovered a property in Central Queensland which the landholder no longer wanted but did not want cleared. The land was on two titles and when Hugo Spooner's father died in 1996 his mother decided to sell the smaller block. Spooner senior's early farming experience had been in East Africa where land clearing did not happen. He was an instinctive conservationist and had farmed carefully, without overgrazing. Hugo Spooner, who still farms the bigger block, maintaining a quarter of the property as a nature refuge, wanted to sell to a conservation organisation.[105] Goonderoo was in the

Summer rains swell the creek that winds through the brigalow on Goonderoo Reserve, Qld.

Photo: Wayne Lawler/Ecopix

Brigalow Belt, a high-priority region identified as seriously threatened and inadequately represented in existing reserves. The Brigalow Belt comprises open forest and woodland communities in which the brigalow (*Acacia harpophylla*), a silvery wattle with a broad canopy, roughly 10–15 metres high, dominates among Belah, Gidgee, Lancewood and Bendee trees. The brigalow shelters patches of semi-evergreen vine thickets with bright green leaves, remnants of the subtropical rainforest vegetation that covered the area millions of years ago. It provides habitat for nationally threatened fauna, including the bridled nailtail wallaby (known as 'flashjack'), northern hairy-nosed wombat, squatter pigeon, brigalow scaly-foot (a legless lizard), and black-throated finch. This type of vegetation was being cleared everywhere, since the soil supporting the ecosystem was suitable for agriculture. Goonderoo still had good brigalow forest because less than 50 hectares had ever been cleared. The Board agreed with the staff recommendation that Bush Heritage should buy it.

Goonderoo was the first property purchased on the basis of scientific assessment, setting a precedent for

South Esk Pine Reserve, Tas

Photo: Wayne Lawler/Ecopix

engagement with academic and scientific communities. At 593 hectares, it was Bush Heritage's first landscape-scale purchase, and demonstrated that the organisation could act decisively and quickly on the national stage. In addition, Goonderoo offered interesting possibilities for scientific and comparative research. The land team built trial sites to compare areas of, for instance, grazing with non-grazing, and burnt sites with unburnt sites. Researchers lived on site, and the place hummed with possibilities. Hugo Spooner, a supportive next door neighbour, looked to the time when rehabilitation of the vegetation would extend the refuge for the nailtail wallaby.[106]

In Tasmania, Bush Heritage purchased the South Esk Pine Reserve, originally bought by the TCT to protect what was the best remaining grove of South Esk Pine in the central east coast region, and to commemorate Ros (Roslyn) Jones, a nineteen-year-old campaigner for Tasmania's wilderness from New South Wales, who had drowned during a rafting trip down the Denison River in 1981. Since TCT did not manage land, it offered the property to Bush Heritage, transferring the property in 1998.[107]

An important milestone for Australian conservation was reached in 1999: the passing of the Environment Protection and Biodiversity Conservation Act. The Act provided 'a legal framework to protect and manage nationally and internationally important flora, fauna, ecological communities and heritage places defined in the Act as matters of national environmental significance'.[108] It would strengthen the hand of Bush Heritage and other conservation organisations throughout the country.

At the start of 1999 Bush Heritage owned eight properties, four of them on the mainland. It had an annual income of $606 000. A quick review of Janice Bird's pie charts shows the breakdown of expenditure: 35 per cent was spent on land purchases, 11 per cent on other land costs, 5 per cent on advertising and fundraising, 17 per cent on wages, 4 per cent on accommodation and travel, 10 per cent on other overheads, with 18 per cent transferred to current assets.[109]

Goonderoo meant managing land specifically to protect a threatened species: the bridled nailtail wallaby.

Photo: Courtesy Steve Parish/Nature Connect

Property acquisition whetted the appetite for more purchases, but financial analysis exposed management issues for current and future properties, and the urgent need to predict what those long-term costs might be. It would keep financial expert Robert Rosen awake at night until he devised a system.[110]

Other discussions centred on the feasibility of remaining in Tasmania. It was clear that the big donations were coming from Melbourne and Sydney, and that a base in either of those cities would facilitate travel to other parts of the country. These and other problems would come to a head in 1999, with two more land purchases and new Board members prompting changes in direction. Meanwhile, personality conflicts within the office were threatening the stability of what was essentially a happy and inclusive work environment.

Internal discord

While Board members were aware of the staff tensions, they had their own challenges in ensuring that governance was sound and that the organisation could function effectively if another major land purchase were made. From 1997 there was also talk on issues that the focus on climate change was raising.

SCIENTISTS HAD LONG BEEN WARNING about the unprecedented build-up of greenhouse gas emissions from the burning of fossil fuels and the impact that a rise of even two degrees Celsius would have on the earth's livability. With the accelerating rise in sea levels, increases in the number and severity of droughts, storms, floods and heatwaves, there were predictions for dire losses of species, failures of ecosystems and food production. The 1997 Kyoto Protocol brought a fundamental change to the way companies viewed their carbon emissions, and Bush Heritage had to consider whether to take advantage of the carbon credit scheme, whereby companies could offset greenhouse gas production against contributions to land conservation. Intuitively it went against the grain, but it could enable them to achieve their goals sooner, a thorny dilemma for an organisation with high ideals. Under what circumstances could Bush Heritage accept donations from companies involved in activities that were detrimental to the environment? Unable to reach a consensus, the Board agreed to develop a policy for the future.

Those on the Board did, however, agree that Bush Heritage urgently needed to develop a broader fundraising capacity and to build a more effective database, to support Janice Bird's and the development program. The difficulty was that finances did not stretch to 'extravagances' that would only show their worth some time after their implementation. Humann and Alexander had already begun a conversation with the Chair and trustees of the JT Reid Charitable Trusts. Their generous three-year gift, starting in 2000, to support such endeavours was a significant moment in the life of Bush Heritage. It occurred at a time when few considered that capacity-building within an organisation merited philanthropic attention. For Humann, it was a vote of confidence, and it began an important relationship with the JT Reid Charitable Trusts that has continued ever since.

On the question of how Bush Heritage saw itself, differences of opinion were emerging. Underlying the Bush Heritage vision was the fact that Australia was a unique continent with 70 per cent of its species endemic, and should be contributing significantly towards the protection of the earth's biodiversity. Bush Heritage was now operating nationally, but with only approximately 2400 hectares of property, could it actually claim to be a leading conservancy agent in Australia? Copley's AWC was surging ahead acquiring land, expanding its donor base, and capturing attention as a brand name in conservation, while the World Wildlife Fund maintained its premier position with its appealing call to save endangered animals. Both were trading heavily on the image of 'wildlife'. In Victoria, TfN had a respectable following. Humann and the Board were keen to reach similar goals, and knew that Bush Heritage needed a major acquisition to announce itself as a significant player in the environment. With her background in management, Mara Bún pressed for developing systems to provide greater transparency and pathways for actions. She wanted to explore the extent to which Bush Heritage intended to be a national organisation, believing the Board and staff lacked the skills to take the organisation to the next level. She was critical of Humann's performance, although she may not have known the extent of his workload that included stewardship of individual donors, donor presentations, involvement on environmental committees at federal and state levels, overseeing management of properties, and travel all over Australia in pursuit of new properties, plus negotiations with vendors and local councils.

Louise Gilfedder had different concerns. A contemporary of Blake's and a protégée of their mutual mentor, Professor Jamie Kirkpatrick, she worried that the Board was not being guided enough by science when making decisions. After her work in the Midlands of Tasmania, a heavily farmed district with vestigial remnant native bush, she knew that the challenges lay in the farmed areas, that pastoral leases contained some of the most endangered ecosystems, because farmers had naturally turned the most fertile land over to grazing and agriculture. It was important, she believed, to develop science-based programs with systems and criteria that could be quantified as well as qualified, where 'you could justify what you were doing and why, set priorities, and balance these with donor satisfaction'.[111] In the 1990s she was establishing long-term relationships with private landholders in the Tasmanian Midlands, researching the ecology of grasslands, woodlands, and associated threatened species. Her brief was to integrate biodiversity into agricultural systems, a primary concern for Kirkpatrick. This represented a major shift in the approach to conservation, but Bush Heritage did not venture down this path at this stage. Paul Sattler would also champion that approach when he joined the Board in 2000 and gradually the concept began to have traction as the Board relinquished its hold on 'gorgeous' landscapes and acknowledged the need to conserve rangelands. As Gilfedder commented: 'We didn't need more wilderness, we had to go for the hotspots'.[112] She advocated taking a unified approach to acquisitions, in line with Blake's systematic assessments. But it was difficult to observe proper targeting principles when a good opportunity arose. As Board members had discovered, a property available for purchase often overtook planned strategies, and this was true for the properties acquired in 1999.

Underlying the Bush Heritage vision was the fact that Australia was a unique continent with 70 per cent of its species endemic, and should be contributing significantly towards the protection of the earth's biodiversity.

A time of transition

In a reflective mood, I turn off the recording, my eyes lingering over the transcript on the screen as I think about the different emphases that voice and the printed word reveal. My interviews have been by Skype, face to face, or through email exchanges. I alternate between the computer screen and the voice, surveying the turbulent first years of Doug Humann's time as CEO, piecing together from those who were there a picture of how it was.

Memory is not as exact or as definitive as I would like; there are shadings that slip across the page, and each memory is coloured by the person talking – who they are, where they were then, where they are now. I know from other research that truth has many faces, including that of the interviewer. While they happened only fifteen years ago, these events are already history; we cannot relive them, but must judge from where we are now.

From small asides and longer comments, I discern a tension in the office, the clash of personalities that could make or break the organisation. I know that it didn't break it – in fact, it probably provided the impetus to face the next big challenge almost on the horizon. But I am curious about its origins, and how it reflects the way office procedures were changing. I wonder, too, about the impact of personality on events. The organisation is still small (only six staff), and there is still a loose exuberance about the interactions, perhaps for the last time. There will be a necessary dampening of personal idiosyncracies as the organisation expands to meet bigger goals. The energy will be harnessed as will the differences, as a more corporate narrative drives the organisation forward, smoothing the rough edges.

The first clues that the closeness of the early years is dissipating are in Nathan Males's account. He is developing visitors' programs for the properties. 'We had this amazing resource of people who contributed money but also contributed skills and enthusiasm, who when they became engaged, understood the aims and direction of the organisation. We all saw the benefits of building up engagement programs and getting people out on field days.' He recalls the office as divided, emotionally and physically: 'Doug's team was the fundraising and administrative staff and the other, Greg's team, the land managers'. Nathan admits to being 'young and naïve'. The early days were 'wonderful. The organisation didn't really know what it was up to, [it] was just finding itself in the world'. The office tension was 'really interesting ... and all good fun'.[113]

For Nathan, 'Blake was an unusual character, quite inspirational, very charismatic'. Perhaps he was too young to sense the charisma of the other. For him, Doug was 'obviously a very solid careful planner and administrator. Doug is in lots of ways a control freak, you had to be to pull the thing together, Greg was too, but in a different way. They were never going to work effectively together. They both contributed enormously, though'.

In 1999, Doug became the CEO rather than the Executive Director. The Board was working towards taking a more strategic role and felt that the position of CEO held more gravitas and was more appropriate for an organisation that was now engaging closely with business leaders and governments. However, the staff were taken unawares and were unsettled by the change. I work through the interviews. Karen Alexander observes that she could have done more to support Doug, when he 'commented to me that things were difficult'. She would fly down to help sort out staff issues after the event instead of anticipating the problems.

The change hit Greg Blake hardest. He had been running the on-the-ground side of Bush Heritage since joining the staff. He now lives in Laos, so we have communicated only through emails. I re-read our email conversations. He remarks that, after a conversation with Karen, there was an understanding among the staff that if they felt they could not support Doug, they should leave. This further inflamed tensions.

Bush Heritage staff in 1999. From left: Jos Bentley (front), Greg Blake, Margaret Graham, Doug Humann, Kerry Green, Janice Bird

Source: Bush Heritage Australia

I have talked with Doug in the quiet of his modest home in Melbourne. He is articulate and thoughtful, and looks back on his fourteen years at the helm of Bush Heritage with justified pride in its progress. His is a strong and measured narrative, relaxed in its recounting, and he alludes lightly to the difficulties, concentrating more on the 'mega picture'. And I note several donors who have remarked on Doug's capacity to inspire them with the Bush Heritage conservation model, with its vision; and on his integrity and honesty that gave them confidence that their money would be well spent. His passion for Bush Heritage is evident, his interest in its wellbeing, long after he has retired as CEO, still very much part of him. When I ask, he confirms that his relationship with Greg was fiery and ultimately unproductive.

I contrast his story with Greg's. Doug already had a wealth of experience when he came to Bush Heritage. He had been a successful teacher, the Executive Director of the VNPA, and he had a wife and children. Greg is younger, perhaps spoilt by his years in charge, and unused to boundaries. Janice Bird, very much in Doug's camp, corroborates this. She remembers that constraints on budgets were ignored, with Greg brushing aside her concerns about managing the finances, going over her head to the Board, where he was met with a sympathetic hearing. After all, he had been on the Board himself, so why not? But they were always short of money. She remembers the 'pleasure [of] talking to people about donating to Bush Heritage because I believed in what Bush Heritage was doing and knew first-hand how tiny our overheads were. I could confidently say that 95 cents of every dollar went into buying or managing land'.[114]

Passion for Bush Heritage drives both Doug and Greg. I read Greg's words: 'I really loved Bush Heritage. It was not a job, it was a calling'. Working at Bush Heritage was 'vibrant, exciting, exhausting, sometimes frustrating but infused with the delight of doing something worthwhile which really worked'. Bush Heritage would become the foundation of his professional career and his philosophical direction for many years, because:

> [it] existed in the wider context of conservation in Australia, as the genesis of an idea and an initiator of a continent-wide movement that took on a range of forms, forcing members of the conservation movement to face the realities of land ownership and management, feral control, the frequent conflict between aesthetics and biodiversity management, issues of grazing, fire and weed control … and many other issues for which people were obliged to take personal responsibility rather than just write letters to the editor. It pushed the 'spiritual' ownership of Australia out of the tame and modified landscapes of the south and into the wild and unforgiving lands of the north and west.[115]

It is the classic fight of the new against the old, but old and new are blurred. What is clear is that there is not room for both; one of them must give up control of the direction and philosophy that underpin Bush Heritage. Greg stands for the passion and relies on raw energy. It has worked up till now, proof being the property assessment guidelines and the management plans for each block. He believes Doug does not understand the scientific underpinning of his work. Doug represents the modern approach, more measured, consultative, disciplined, working more cautiously with donors to build long-term relationships and develop the reach of the organisation. Inevitably, Greg raises the issue of staff salaries and better working conditions. Doug will not recommend this to the Board; his sights are set on expanding the horizons of Bush Heritage. He sees Greg as unreliable and divisive. For Greg, the issue is more complicated. He recognises now that he was 'naïve, overly energetic and communicated badly'. He can see that the ongoing conflict, without possibilities of negotiation or compromise, will harm what he loves; it will be unproductive and could destroy the gains they have already made. He wants the organisation to consider alternatives, and work in a collaborative way, rather than melt into a hierarchical structure. He can either remain silent and passively accept the situation, or leave. He resigns. He will work for Queensland TfN, TLC and AWC before forming his own design and management company and eventually moving to Laos.

I had envisaged that Bush Heritage's development would entail some painful steps, but the extent of the personality clash is a surprise. I query this with Greg, and see on my screen his acknowledgment. He left disenchanted, but with hindsight recognises that: 'At the time, I did not understand the normal birth pains that organisations went through in their transition between things that are driven by the passion and commitment of individuals to larger things that are maintained and directed by process and, in retrospect, necessary regulation. The ego is threatened by such transactions and in my case I had difficulty with that'. Ironically, one of his last tasks for Bush Heritage was to assess the property that would be the making of the organisation, and would bring Doug truly into his role as a great CEO.

PART 2

As more donors pledged their support for the Australian Bush Heritage Fund, there was a pressing need to underpin the passion of its staff and volunteers with a stability that would attract new people to the organisation. Having made small forays onto the mainland, the Fund's first big purchase in Queensland committed it to a bigger program and the decision to relocate to Melbourne.

Making the vision a reality

A turning point

For several reasons that nobody could have predicted at the start of the year, 1999 was pivotal in Bush Heritage's development. It was the year that the bequest program, begun as a consequence of informal talks between President Karen Alexander and Roewen Wishart, then working for Oxfam, began to yield positive results.

THE PROGRAM WOULD CONTINUE to grow, as more and more donors confirmed their intention to leave a bequest to Bush Heritage, and this source of income would become a vital part of Bush Heritage's program. The year 1999 was also important for conservation in Tasmania. The government established a Protected Areas on Private Land Program, offering voluntary conservation covenants to farmers and other landholders. This was the result of work by Professor Kirkpatrick, assisted by Louise Gilfedder, as a response to national conservation priorities. The program used a range of incentive programs to build a comprehensive, adequate and representative (CAR) reserve system in Tasmania's non-forested areas that would support the protection of rare, vulnerable and endangered non-forest vegetation. Although this would later be significant for Bush Heritage, the new government policy did not affect its current direction, since it concerned grassy farmlands that were not yet part of Bush Heritage's strategy.

In 1999 the Board also approved the acquisition of several properties. The first of these properties, Tarcutta Hills – 432 hectares in New South Wales, 130 kilometres west of Canberra – contains the last fragile remnant of grassy white box woodland in good condition in Australia. Similar woodlands once clothed the hills of Victoria and New South Wales; this was the country of lost children in early settler days, the alien bush that bred fear in the hearts of those used to European countryside, even as it drew them in. Now immortalised in literature and art, the real thing has all but disappeared. The name, 'white box gum', is misleading, since the bark is grey and coarsely fibrous around the base and on the major branches, with only the thin ends of branches sheathed in smooth white bark, like glowing, unblemished skin. From a distance the drooping foliage is grey, too. The light understorey includes several different native grasses that are habitat for numerous animal species, among them the uncommon yellow-footed antechinus. The occasional massive old tree provides hollows for nesting birds, small mammals, and the nationally endangered turquoise parrot and swift parrot. The white box grows in the company of several other species of eucalypt: Blakely's red gum, yellow box, and several kinds of ironbark – red trunks against black, cream and grey ones, randomly and widely spaced, the patterning never the same, but over and over again, impenetrable because there is no sky between the trees, only more trees, and a stillness under the variegated greenish-grey canopy. Deceptively easy to walk through, the native grasses create an illusion of openness; it is easy to follow the contours of the land, down through shallow stony dips, and up the next ridge, in search of something more, the heart of the forest perhaps, that is never quite attainable.

In the early 1900s the property was granted to the NSW government railways department as a source for wooden sleepers to support railway tracks, since ironbark is resilient and termite-resistant, the toughest timber in the country. Sleepers were cut from the property for many decades. After the federal government had identified white box woodland vegetation type as one of the most seriously threatened and valuable ecosystems in Australia, the property was passed to the NSW forestry department and then sold at auction to the owner from whom Bush

Tarcutta Hills Reserve, NSW: a last remnant of white box woodland

Photo: Wayne Lawler/Ecopix

Louise Sylvan, with Doug Humann (left) and Greg Blake at Burrin Burrin

Source: Bush Heritage Australia

Heritage purchased it. Ironically, an appropriate bureaucratic process could have transferred it directly to the National Parks and Wildlife Service for permanent protection since it had always been recognised as a top conservation priority. When Bush Heritage bought it, the land had never been grazed, and was only partly cleared, the owner having maintained the property as a nature reserve. For land manager Greg Blake, it was a magical place: 'One morning I went outside just on dawn and the chorus of birdsong was deafening. I thought, What a fantastic place, so full of life … and then it struck me, with real intensity, that once, not long ago, the entire landscape was like that, and now [it] is mostly silent'.[1]

The next property purchased in 1999, Burrin Burrin, once belonged to deep green environmentalist and philosopher, Richard Routley (later Richard 'Sylvan'),

Burrin Burrin Reserve, NSW, is mountain country with mists and ferny valleys.

Photo: Wayne Lawler/Ecopix

who bought a number of properties for conservation purposes with his first wife, activist and ecofeminist intellectual, Val Plumwood. They were radical philosophers, questioning man's relationship with nature. They warned about the 'environmental crisis' in their 1970s book, *The Fight for the Forests*, one of several influential books alerting the public to Australia's policy of converting its hardwood forests into woodchips.[2] Routley and Plumwood's twenty-year partnership and subsequent independent work included a total rethink of humanity's ethical relationship to nature with their desire to live simply in the bush. Burrin Burrin's remoteness attracted Routley, and he found a kindred spirit in the owner of the neighbouring property, Sharewater. Author Julian Davies had bought his property in order to maintain the surrounding bush, and the two men explored their land together. The properties extend inland from the top of the southern NSW escarpment and rise from wet rainforest valleys up to 1000 metres, and are home to many types of vegetation, fourteen different types of eucalypt and numerous other plant and animal species.

Alex Griffiths and friends at Currumbin Wildlife Santuary

Source: Bush Heritage Australia

Burrin Burrin's 114 hectares of intensely folded country on the eastern side of the Great Divide shelter little ecosystems tucked into pockets in the rocks and steep gorges, that are home to sugar gliders, superb lyrebirds and powerful owls.

Richard Sylvan died in 1996, leaving Burrin Burrin in his will to the Australian Conservation Foundation (ACF), although he had latterly wanted Bush Heritage to have it. ACF did not want to own property, and transferred it to Bush Heritage for the sum of one dollar, since Routley's will required that ACF make a profit from the sale. It was an easy addition to the Bush Heritage portfolio, a small wilderness posing no great management problems, with its neighbour (Davies) willing to maintain a watch over the property.

Bush Heritage's next acquisition was Currumbin Valley Reserve, 4 hectares in the hinterland of Queensland's Gold Coast, that acts as a buffer for the adjacent Nicoll Scrub National Park, protecting a rare remnant of rainforest in south-east Queensland. It is habitat for the vulnerable Richmond birdwing butterfly and long-nosed bandicoots, green catbirds, scrub turkeys, and several honeyeater species. Although logged, it still protects rare plants, such as the nationally endangered sweet myrtle and several nationally vulnerable species of macadamia. The property was donated to Bush Heritage by the Wildlife Preservation Society of Queensland according to the wishes of the previous owner, Dr Alex Griffiths, a birdwatcher passionate about conservation.

In 1999 Bush Heritage also increased its holding at Kojonup in Western Australia with a small parcel of land, and in May 2000 Humann formalised the related covenanting arrangements in partnership with the WA National Trust of Australia. The celebration generated interest from other landholders in private land conservation and covenanting. As the Board did not intend at this time to develop a presence in Western Australia, Board member Margaret Robertson concentrated on the Kojonup property herself. Her views point to directions that Bush Heritage would take in the near future:

> My priorities were different to other Board members. I was working remotely, and the Kojonup block was surrounded by farmlands and people wanting to do hands-on things to conserve the property. We had productive sessions on Kojonup ... driven by the people here, not just for conservation, but for community outcomes. Angela Sanders [now a Bush Heritage ecologist in Western Australia] was planning the Kojonup anniversary the following year, and we brought out people with interests and skills, for example keen bird lovers, and other knowledgeable people, who contributed their expertise. It wasn't Bush

Heritage's direction then, but I felt there was potential for something like this to grow bigger.[3]

The various acquisitions in 1999 generated an increase in donations. But Bush Heritage was still buying 'museum pieces', even though most were adjacent to a bigger park or reserve, so that environmentally they were more valuable than their size might suggest. The organisation's growing public profile and the increased donor support that came with each property helped develop the Board's confidence in expansion. Should they consider a bigger purchase?

President Karen Alexander certainly felt so. For three years, she had overseen Bush Heritage's transition from a local organisation to one operating on a national scale with a more professional approach that made it more open to innovation and new ideas. Staff reviews that she had instigated painted a nuanced picture: the growth in buying and managing properties revealed a successful transition to better management practices, but also pointed to staff struggling under the stress of rapid growth, and on derisory salaries, with much of the physical workload being undertaken by a growing band of volunteers. Volunteers were an important part of developing a holistic program for each new acquisition, but Humann foresaw implications for the health and safety aspects of the organisation without a more detailed strategy to implement the changes. The demands of fundraising were putting pressure on staff in the cramped office space, while Humann was shouldering an almost impossible workload that in turn impacted on his staff. His 'amazing ability to set priorities' kept staff on track but could not alleviate distress when there was too much to do. This all came at the cost of staff burnout and the loss of Humann's valuable time from his central role of energising and inspiring people about Bush Heritage's conservation vision.[4] He was aware, too, that he did not have unanimous support from the Board.[5]

Two Board members were not confident that Doug Humann was the right person for the job, believing a high-profile, business-oriented person would be more appropriate. This view may have been too 'high-flying' for Bush Heritage 's low-key ambience at that stage. It failed, too, to recognise what Humann, without sufficient business or economics experience, had achieved under difficult circumstances.[6] Humann suffered from the implied and often overt criticism over several years, while believing that directors' input was important, that 'we learnt a lot from their different knowledge of business'.[7] Was it their 'different knowledge' that discontinued the practice of senior staff contributing at Board meetings, with Humann presenting all the necessary information? For a time, the staff attended as silent members, until it became impossible for Humann to answer the full range of technical questions that arose.

Grove of Bangalow or Piccabeen palms, Currumbin Valley Reserve, Qld

Photo: Wayne Lawler/Ecopix

Alexander was troubled and disappointed that Humann did not have the support of all the Board members. His environmental credentials were beyond doubt, as were his empathy and clarity when engaging with benefactors, who trusted his judgment and honesty.[8] But in 1999 two Board members refused to endorse renewal of his contract for three years, believing he lacked that indefinable sparkle. Treasurer Mara Bún saw him as too low-key for the role. For Penny Figgis, herself a vibrant extrovert, he didn't have the dynamic profile she expected from a CEO.[9] This discontent continued as background noise for some time. Alexander proposed the compromise of a two-year contract renewal, which was supported by the two directors and at least kept Humann in his position. Meanwhile this setback faded as the Board confronted its biggest challenge.

Bush Heritage was not yet in this league, and it seemed an impossible ask ... But out of the blue, in July 1999, Humann received a phone call.

Following the successful purchase of Goonderoo in Central Queensland, Greg Blake and Queensland-based ecologist Rod Fensham set about identifying other options for purchase in Queensland. After identifying four possible properties with significant conservation values that would have been under threat if bought for ongoing pastoral uses, they decided to put a proposal to purchase Carnarvon Station to the Board. Since issues involving Aboriginal land ownership were already on the agenda by this time, Humann checked with the Indigenous Land Corporation (ILC) to see if they were interested in making the purchase, and worked with the Commonwealth to find out whether there would be any native title implications if Bush Heritage purchased the land.

Carnarvon Station, 59 000 hectares in Central Queensland, 200 kilometres south of Emerald, abutting Queensland's spectacular Carnarvon Gorge National Park, was one of the few remaining strongholds for woodland species that were declining across eastern Australia. Rugged sandstone hills, narrow valley floors and high escarpments offered natural protection to over 200 animal species, including geckos, gliders, honeyeaters, the tiny narrow-nosed planigale, the common dunnart, as well as hundreds of plant species. The lowland woodlands and bluegrass grasslands were important buffers for the rugged ranges of the neighbouring national park. The property had been farmed since the 1860s, first with sheep and later with cattle; it had had many owners and several changes to its boundaries.[10] When Blake and Fensham first saw the property, it was stocked with 4000 cattle, with crops of sorghum, oats and barley on the plains. Vegetation had been cleared, and regrowth destroyed; the property was degraded by continuous agricultural use over many years. But the bare bones of what it could be, if restored and managed, were discernible.

The assessing staff could also see the possibility of earning income at Carnarvon. The invasive, highly flammable buffel grass (introduced from South Africa) thrives and spreads when burned, and had eradicated most of the native bluegrass vegetation. Buffel grass is vulnerable to heavy grazing, so that selectively grazing it with cattle could eradicate it, allowing the re-establishment of native perennial grasses, and the sale of the cattle at the end of the program, a controversial solution for those Bush Heritage donors who felt there should be no grazing on Bush Heritage properties. The current owners wanted to sell and, because the beef market had just crashed, Humann was able to negotiate his way from the original asking price of more than $2 million through to a purchase price of $1.5 million.

It was a huge price tag. Bush Heritage was not yet in this league, and it seemed an impossible ask, from both management and fundraising perspectives. But out of the blue, in July 1999, Humann received a phone call. He flew to Melbourne to meet a young woman whom he knew from his days at the Victorian National Parks Association (VNPA). They sat at a formica-topped table in a cafe in inner-city Fitzroy, and the young woman asked Humann which he would prefer: a cheque for $1.1 million or for

$1.2 million? Humann opted for $1.2 million, whereupon she pulled a chequebook from her bag and wrote the cheque, with the condition that her donation would remain anonymous. They had a cup of coffee and afterwards the donor rode off down Brunswick Street on her bicycle and Humann, clutching the cheque, took the bus to his overnight accommodation. It was an iconic moment for Bush Heritage, remembered by staff as the turning point. When asked why she made the donation, the donor explained that the money was inherited, not really hers. She had researched various organisations to find a suitable recipient, and when she read the Bush Heritage literature she knew this was the right thing to do, that the money would protect the Australian bush she had loved from childhood.[11] One selection criteria for Bush Heritage had been the property's eligibility for federal government National Reserve System (NRS) funding of two dollars for every one dollar raised privately. That cheque made the difference and purchasing Carnarvon Station was now possible. The deadline for the sale was approaching; a decision had to be made. Early in 2000 the project came to the Board. Alexander was confident:

> Assessments of the property's ecological, environmental and biodiversity values against our own criteria had been done by Greg and his staff; the fundraising needs were understood; purchasing strategies, assessments and management for the first three years had been modelled. There were problems:

Carnarvon Station Reserve, Qld: a spectacular landscape of woodlands, rare grasslands and escarpments

Photo: Wayne Lawler/Ecopix

feral pigs, goats, horses and weeds as starters. We understood the costs – a huge leap in management and fundraising, made more difficult by the distance of the property and its awkward access. The neighbours were prickly, and, although we tried, we didn't really understand the reactions of the Aboriginal landowners. They seemed uncertain, if not hostile, but there wasn't time to tie everything down; we had to decide. We had done research from practically every angle so we understood the challenges. We had been looking at this stuff for a whole year, there was nothing more we could do – it was decision-making time.[12]

The Treasurer, Mara Bún, said 'No'. She felt the purchase was too risky; the strain on management and fundraising capacities would be catastrophic. For a moment it seemed as if all the assessment work might have been for nothing. But others disagreed. Robert Rosen, a former treasurer, was positive. He believed the alternative of doing nothing was a greater risk. The Board decided to buy, with Bún

Mixed forest at Carnarvon Station Reserve provides habitat for many species.

Photo: Wayne Lawler/Ecopix

dissenting on the decision for reasons of financial risk and management capacity. She was 'completely and utterly terrified by the idea, and I resigned at that meeting. Then I changed my mind and returned. In the end it was a good decision; you just have to take the plunge, make the commitment to succeed, and then you will'.[13]

Carnarvon Station was high on the Queensland government acquisition wish list, and Bush Heritage gained credibility by buying it. Alexander was elated. 'We were going into a new phase and we were ready.' Bob Brown's original risk-taking gesture had set a precedent. Alexander hoped the purchase would encourage interest from the wider community, and provide impetus for internal change within the organisation.[14]

Because Bush Heritage had to change, if it were to survive. The Carnarvon purchase brought clarity for the always pragmatic Alexander. With her strategic skills, developed and honed through countless environmental campaigns, she recognised that Bush Heritage needed to grow (and the purchase of Carnarvon Station implied a new level of growth) and for that it needed a president with a new vision, strong leadership and negotiating skills, and a high public profile. It needed someone who was media savvy and had access to government departments, business and community advocates, and was comfortable on centre stage, someone who would push Bush Heritage into the limelight. Alexander did not see herself as this person. She preferred to lead through suggestion, consensus and supporting others' initiatives, while remaining in the background. As well, Bush Heritage had to consider moving to the mainland. This was not a new idea; relocation had been bandied about since 1993, because Sydney and Melbourne held the key to fundraising potential. Alexander had discussed the move with Bob Brown, believing it was essential for the organisation, but her strong personal relationships with the founders of Bush Heritage made it stressful to negotiate this move.

Alexander's thoughts turned to her long-time friend and colleague, Phillip Toyne. Before she became a freelance environmentalist, she had directed campaigns for the ACF when Toyne was its Executive Director. His broad environmental experience, his contacts in government and with business, combined with a personality that was both attractive and tough, would be an asset. She knew Toyne and Humann would like each other, and that Toyne could guide and mentor Humann through this period.

Bush Heritage had to consider moving to the mainland. This was not a new idea; relocation had been bandied about since 1993, because Sydney and Melbourne held the key to fundraising potential.

After leaving the ACF in 1992, Toyne spent two years as a visiting fellow at the Australian National University in Canberra, at the Centre for Environmental Law. His teaching involved research into the corporate history of the environmental movement, which was misunderstood or unknown, partly because of the high turnover of personnel, especially of young people. His research resulted in a book, *The Reluctant Nation*, and a radio series for the Australian Broadcasting Commission. He was appointed deputy secretary in the federal Department of Environment, a position he held for three years, before opting out of the public service. He and his second wife, Molly, another committed environmentalist, were living at Gundaroo, a small country town north of Canberra, and had established a company, Ecofutures, to undertake a range of environmental activities. They wanted to earn a living in a way that gave them more family time, and space to volunteer on environmental issues that were important to them.

All this was known to Alexander. Some time between April and the end of October 2000, after discussion with the Board, she phoned Toyne and asked him to consider becoming President of Bush Heritage.

Activist, lawyer, teacher, advocate

Phillip Toyne's initial reaction was cautious. Alexander had often spoken about Bush Heritage, and he knew she found the organisation inspiring. On one occasion when he attended a donor meeting, he saw a parochial body operating on a small scale in Tasmania, with little presence elsewhere. He wondered why Alexander thought Bush Heritage might interest him, and why she thought he would be good for it.

MUCH OF TOYNE'S HANDS-ON environmental experience had been of an activist nature, leading campaigns to stop institutions destroying the environment. He arrived as Executive Director at ACF in 1986 at the perfect time and to a position tailor-made for his personality. He was in his element negotiating the politics during a period when the general public was awakening to environmental issues.

Under Toyne, the ACF took up other campaigns in the 1980s and 1990s: expanding the boundaries of the Kakadu National Park, and forcing an extension of the World Heritage Area while protecting Coronation Hill from mining. He drove the decision to block international moves to allow mining in Antarctica, and to list the wet tropic rainforests of North Queensland as World Heritage sites. Before joining the ACF, he was instrumental, too, in the handover of Uluru Kata Tjuta National Park to its Traditional Owners. A towering figure in those turbulent times, he was a symbol of 'green' activism for many Australians.

Although Bob Brown was a friend, Bush Heritage was not on Toyne's personal radar, and he did not fully understand its rationale. But his curiosity was piqued by Alexander's insistence that he would never be bored working with Bush Heritage. She explained what she wanted to achieve, and that she thought he could take the organisation to the next level. He trusted her, and agreed to meet the staff and the Board, and to consider the idea with the proviso that he would only be interested if he liked Doug Humann and believed he was a capable CEO.

Looking back, Toyne remembered:

> I had lunch with them [the staff] and I was very taken by their vision for what might be. And I was certainly taken by their energy and enthusiasm. I could see lots of things that needed to be fixed. That was sufficient for me to feel it was something I should put some effort into.[15]

Phillip Toyne, President of Bush Heritage (2001–08)

Source: Bush Heritage Australia

Later he met the Board, and there were familiar faces: 'It was a little greeny board, that is the only way you could describe it. Most of them had been recruited on the basis of their environmental activist credentials'.[16] Toyne saw that the Board members were strongly focused on tall trees and old forests, and still very much involved in the everyday running of the organisation. He also sensed doubts among some members about Humann, and understood that they had been 'traumatised' by the decision to buy Carnarvon Station.

Through the prism of his inner restlessness and a driving personal ambition to succeed in areas that mattered, Toyne recognised the problems for Bush Heritage, but also saw its big heart, and its potential. He told Alexander he would accept the position if it were offered. He was invited to the November Board meeting in 2000 as an observer. At the meeting Alexander indicated her decision to step down as President, nominating Toyne as her successor. With no other nominations, there was unanimous agreement for Toyne to assume the presidency at the next Board meeting in March 2001.

Through the prism of his inner restlessness and a driving personal ambition to succeed in areas that mattered, Toyne recognised the problems for Bush Heritage, but also saw its big heart, and its potential.

When Toyne became President he was fifty-three, a tall man who moved with natural grace, and looked younger than his years. Perhaps that was the larrikin in his soul, the cheeky, slightly weatherbeaten Australian look about him, the straight brown hair that flopped across his forehead, and grey-blue eyes that held an intensity just as well as a lazy smile. His innate charisma and laid-back demeanour masked a gritty strength of purpose, and a coiled, nervous energy. His maturity had been gained during a varied and disjointed work history, and as he instinctively rose to each challenge, his outlook had expanded to encompass the idea of a more inclusive Australia, and the desire to protect the natural environment. These two strands came together during his presidency, and helped refine the Bush Heritage vision.

Toyne's early experiences were significant in shaping the direction he took with Bush Heritage. He began studying Law in 1964 at the University of Melbourne, during the period of political and social upheaval that swept through universities in the Western world and in Australia against the backdrop of growing disquiet about the Vietnam War. Many, including Toyne, were convinced they could make the world a better place. He had thought Law was the 'vehicle for social change and justice and a support for the poor and disadvantaged', but this belief was soon 'shattered'.[17] After graduating he worked in the field of workers' compensation, coming to see it as a cynical exchange between insurance lawyers and TAC lawyers. He 'didn't want to be part of that system', and left to travel around Australia, aged twenty-three, just three months shy of completing his articled year which would have enabled him to practise as a lawyer. This impulsive move would return to bite him.

His epiphany came at a dusty roadhouse stop. Some Aboriginal people were sitting in the shade beside a ramshackle collection of humpies. He greeted them but they did not respond. It dawned on him that they were the first Australians, and he 'couldn't communicate with them'.[18] They were living in terrible conditions which seemed unjust, and he became aware of the huge gulf of cultural understanding that existed between them. He was not sure why, but at that moment he decided to become a teacher. With a sudden flash of insight, he realised that influencing children's ways of thinking was the way to change society.

Toyne's understanding of the significance of land to Aboriginal people was pivotal to the way he sought to develop these partnerships.

Back in Melbourne, the La Trobe University teaching program was radically different, and Dr Doug White was a powerful influence on the young Toyne. The course encouraged teachers to bring influences beyond the standard curriculum into schools, and to consider various educational outcomes. Halfway through the course, Toyne made another impulsive decision: to teach in the outback. This was not difficult, the only requirement being a name on a form. He was assigned to Haasts Bluff in the Western Desert, 230 kilometres west of Alice Springs.

For the city-bred Toyne, his two years as the only teacher in a remote Aboriginal community took him to the edge of cultural difference and isolation. With no previous teaching experience and no mentor, Toyne found his own way to teach Aboriginal children of all ages in a class that numbered from five to fifty, depending on the day. They gravitated to the young man who wanted to learn their language and immerse himself in Aboriginal life and culture, despite disapproval from the white community. The station manager considered Toyne's open-gate policy, particularly when allowing the children a weekly hair wash with Education Department shampoo under the garden hose in his yard, as 'inviting trouble', saying that 'it would be his problem' if anything happened.[19] Toyne adapted his teaching to the environment, engaging students with active learning situations. But, as he remarked later, it was he who learnt the most.

His success can be gauged by the way former students kept in touch over the years. When he returned to visit the community many years later, in 2011, they travelled great distances to greet him. The relationships forged at Haasts Bluff would inform all Toyne's dealings with Aboriginal people, particularly when, as President of Bush Heritage, he helped develop productive partnerships with Aboriginal landowners during the early 2000s.

Toyne's understanding of the significance of land to Aboriginal people was pivotal to the way he sought to develop these partnerships. He learnt that land was 'the central and inseparable part of their being'.[20] His description of the songs or *tjukurpa* of the time of creation predates Bruce Chatwin's theories of songlines and gives background to the journey that John Bradley, a young teacher working with Aboriginal children in Borroloola in the 1980s, underwent discovering the sacred songs of the Yanyuwa people in the Gulf of Carpentaria.[21] Like Chatwin and Bradley, Toyne recognised that creation songs were the 'key to rich, complex and intricately related knowledge systems, where every part of the landscape is named and is related to the past and present social and cultural life of the people, and provides the "rational and moral" order to their existence'.[22] A subtle and complex concept, it has to be recognised in the context of Toyne's desire to integrate Bush Heritage conservation work with that of the Traditional Owners of the land.

When Gough Whitlam and the Labor Party won government in 1972, Toyne was thrust into the politics of land rights issues. Whitlam, having promised to legislate for Aborginal land rights in the Northern Territory, set up the Woodward Royal Commission in 1973 to explore ways Aboriginal land rights could be recognised. And Toyne, the almost-lawyer, discussed what this meant with his desert community. He was distressed to have to tell the people that they did not, in fact, own the country they considered theirs, and that they had no legal rights against someone wishing to use their land for other purposes. Once again, the spectre of injustice for vulnerable people was raised. Toyne was therefore open to an invitation from the newly established legal aid service in Alice Springs to play a role as a lawyer because of his experience in Aboriginal communities. He jumped at the opportunity, even though it meant returning to Melbourne to complete another nine months in a law firm to obtain his articles so that he could be admitted to the Bar. Once again, he was plunging into new territory, 'biting off more than I could chew'.[23]

Returning to Alice Springs, he spent two years defending young Aboriginal people in the justice system, before being coopted onto the Central Land Council. He began working on land claims after the NT Land Rights Act was established, before becoming the lawyer for the Pitjantjatjara and Ngaanyatjarra peoples, who had land claims in the Northern Territory, Central Australia and South Australia. He found himself dealing with the SA Premier Don Dunstan when Dunstan legislated for a new SA Land Rights Act to replace the version in place before Whitlam's 1973 Act.

The land claim application for the inclusion of the north-west corner of the Maralinga atomic bomb testing area in the Maralinga claim introduced Toyne to the complexities of land conservation, from both the Aboriginal and Western points of view. Although not obviously remarkable, the reserve was highly prized by scientists and conservationists because of its comparatively pristine state and its endemic flora and fauna species. Progressive conservationists supported the Aboriginal application, but others believed that conservation areas should never be Aboriginal-owned, because they would not be adequately protected, and because the land should belong to everyone. Later, having been involved as a lawyer in the 1985 Uluru handover to the Traditional Owners, Toyne saw it as a successful attempt to bring conservation and Aboriginal ownership into positive co-existence. But working for the inclusion of the Maralinga site was his first real encounter with the conservation movement, and he never forgot those insights. They would underpin his future career choices, particularly when he was considering the presidency of Bush Heritage. As he intuitively followed his passions, he was drawing closer, without perhaps realising it, to the environmental movement. By 1986, when he accepted the presidency of the ACF, he knew that protection of the land was his paramount goal.

Phillip Toyne with Traditional Owner Matilda House at the launch of Scottsdale Reserve, NSW, 2007

Photo: Eddie Misic

In the ACF he developed management skills and a public profile, and the contacts and friendships from those years remained as sources of support. Two relationships in particular, with Peter Garrett and Rick Farley, would be important for Toyne's later work with Bush Heritage. Garrett, the former lead singer with the group Midnight Oil, which he joined in 1973, had a long history of political activism. He was President of the ACF from 1989 to 1993, a time that overlapped with Toyne's stint as its Executive Director, and again from 1998 to 2004. Their association had implications when, in 2007, Labor Prime Minister Kevin Rudd appointed Garrett as Minister for the Environment, Heritage and the Arts, a title that changed significantly in 2010 to include 'Environment Protection'. Doug Humann commented that public servants in the ministry 'used to worry when they knew Phillip was talking to Peter. They never knew what Phillip would suggest, because he always had heaps of ideas'.[24]

During Toyne's ACF days, Rick Farley was President of the National Farmers Federation (NFF). Theirs was an unlikely alliance, greenies and farmers being traditionally at odds over pretty much everything, particularly land use. But Toyne and Farley joined forces because of the continuing degradation of pastoral land, and the need to consider alternative land management policies that would benefit their constituents' goals – in Farley's case, more resources for farmers, in Toyne's, the possibility of instigating greater conservation activity in the bush than environmentalists could achieve by themselves.[25] Both saw advantages for their organisations in Victorian Premier Joan Kirner's Landcare program. Toyne was concerned about arid lands failing because of bad farming practices, and Farley, coming from a different angle, saw in Kirner's model ways for farmers to improve their pastures while gaining environmental credentials. Throughout 1988 Toyne, Farley and others worked on a joint Landcare submission for 'a major integrated program which [would] incorporate a local, cooperative self-help approach, the development of individual property plans which [would be] integrated into district plans, and increased assistance of State and Local Governments within an overall framework of Commonwealth policy guidance'.[26]

Few politicians could afford to ignore an NFF–ACF partnership pushing Landcare, since it reflected a growing concern about the environment. Toyne and Farley were successful in presenting their case to the federal government. Labor Prime Minister Bob Hawke, in his 1989 address, *Our Country, Our Future*, announced the One Billion Trees program, and a national Landcare Decade, to start in 1990. He appointed Greening Australia (GA) to carry out the program. Landcare's activities ranged from soil conservation, management of erosion and salinity, sustainable farm practices, restoration of native habitats, revegetation and control of weeds and pests to the development of local natural resource management skills and knowledge, and to planned publicity and teaching programs to spread the message. GA identified a radical direct-seeding system technology as the means to achieve the target of a billion new trees, holding a direct-seeding conference, 'Sowing the Seeds', in 1990. The conference produced technical papers that set new benchmarks for the 300 experts in Australia, and gave hope to many Landcare groups that their objectives could be achieved. At the point where green activism met conservative farmer, care of the land became everyone's responsibility. Toyne and Farley won plaudits for their cooperation in bringing the project to fruition, and their public reputations, as well as those of their organisations, were enhanced.

The two remained friends and colleagues until Farley's untimely death in 2006. Toyne suggested they were both 'very self-confident alpha males'.[27] Jane Elix from the Women's Electoral Lobby, who worked with them on the Landcare project, believed they 'shared a front of arrogant charisma', and the ability to take intuitive leaps in their planning, but were 'not so fabulous at following through'.[28] As soon as he could, Toyne invited Farley onto the Board at Bush Heritage, where his energy and out-of-the-square thinking were invaluable.

Stepping up the pace

Before his first Board meeting, Toyne talked with Board members and staff, particularly Doug Humann. When exactly he decided Humann was the right man for the job is not clear. But Toyne spent his first weekend in Hobart with Humann and his family, enjoying the convivial ambience, so that what became a lasting friendship between the two men might have begun at that time.

HUMANN, HIMSELF, WAS NOT SURE about Toyne; they had worked together briefly when Toyne was at ACF and Humann at VNPA, and he knew Toyne was a risk-taker, and much more flamboyant than he was. Because of Toyne's legendary status in the conservation world, Humann was doubtful that Bush Heritage was big enough to engage his attention. Perhaps both were surprised at the commitment Toyne brought to the job from the outset.

Toyne's first task was to resolve the issue of Humann's continuation as CEO. He also recognised the Board's lack of experience in marketing and fundraising, and knew that an outreach to a broader audience of potential supporters, and a higher level of professionalism in all areas, were necessary. Central to this were the skills required from the CEO. Given Bush Heritage's ambitious plans to expand, Toyne queried whether they were expecting too much of Humann, pointing to major structural problems within the organisation. As Toyne was 'transparently honest', Humann accepted Toyne's expression of confidence in him that became apparent at Board meetings. Once he knew that he had Toyne's support and that Toyne was defending him against the background hum of discontent, he worked more confidently with the Board.[29] His relationship with Toyne would unlock the potential within Bush Heritage.

Toyne's presidency began at the Board meeting in March 2001. A reading of the meeting minutes reveals a sweeping new energy. Toyne quickly cut to the core of the business with questions and recommendations that showed he had done his homework and was ready to step up the pace. Even though Bush Heritage was not an advocacy group, at Toyne's urging the Board agreed that Humann should take a position to the government on three key issues: the NRS, tax incentives for investors in conservation, and laws on pastoral leases and covenanting. Doubts about the future of the NRS were raised, and Toyne, knowing how vital this support was to Bush Heritage, recommended that Humann put more time into future interactions with the responsible Commonwealth department, and into further discussions on tax reform with the relevant people in Canberra. Humann was already working with Carl Binning from CSIRO and Michael Looker, who at the time was working for Trust for Nature (TfN), on ways to improve investor opportunities in conservation, which clearly had to be linked with tax incentives. They had developed a paper with Philip Tabas, a lawyer in Boston, along the lines of the US tax innovations. This paper was duly presented to the Australian federal Treasurer, Peter Costello, who approved reforms that made it easier to put covenants on land. Costello also agreed to extend the cultural gift program to nature reserves, which meant that people bequeathing or gifting property, cars or artworks to an environmental organisation could now claim a tax deduction. The organisation could sell the item to recover the money for specific use in its programs, and the money would attract a tax deduction that could be spread over four or five years. Toyne also encouraged Humann to lobby for uniform laws across Australia for pastoral leases, particularly regarding covenanting arrangements. While Victoria and New South Wales already had covenanting conditions in place, and progress in Queensland and South Australia was beginning, there were still difficulties in Western Australia.

Toyne's first Board meeting examined Bush Heritage's position within the environment field in Australia, and Toyne, concerned over Bush Heritage's failure to develop a broader public awareness, suggested that they make a major push to improve this situation. He also wanted to revisit the constitution, asking Allens Arthur Robinson, Bush Heritage's legal advisers, to undertake preliminary preparation of a new constitution for presentation at the next Board meeting and to the 2001 Annual General Meeting. Further, he wanted to extend the pro bono relationship that Bush Heritage had with the law firm to others providing services; for instance, to encourage Bush Heritage's accountants, advertising agencies and printers to work in partnership with Bush Heritage. One such partnership that developed out of Humann's friendship with a donor, Graham Turner, would assist Bush Heritage in a very practical way: Turner co-founded the successful travel agency Flight Centre, and his generous annual donation was committed to cover travel for Bush Heritage personnel. Toyne also took up Alexander's draft of Key Performance Indicators for the CEO position, proposing that he and Humann, with the help of Board member Brian Whelan, liaise on some amendments and redraft the document. He finished the meeting by offering his property Gunderoo as the venue for the July Board meeting, and indicated his availablity for day-long meetings if needed.

Toyne was prepared to take risks, and his enthusiasm and confidence carried people with him. His ability to respond to unexpected situations, to adapt to new possibilities, was striking.

The minutes convey Toyne's confidence and decisiveness, the spirit of a new presidency. While Alexander had been ideal for the first years, the Toyne/Humann era pushed the original vision to new heights. Like Bob Brown, Toyne was prepared to take risks, and his enthusiasm and confidence carried people with him. His ability to respond to unexpected situations, to adapt to new possibilities, was striking. But the more important truth about Toyne's presidency was that he knew where he wanted to take Bush Heritage, and that changes were necessary. He was committed to the goal and, even if he did not know exactly how to achieve it, he had the confidence, energy and arrogance to make it happen.

When Toyne arrived in 2001, Bush Heritage had just taken possession of Carnarvon Station. Marketer-fundraiser Charlotte Grimshaw was preparing for the launch of the property at Brisbane's Botanical Gardens, working to ensure wide coverage in the press, and that donors participated in the celebrations. Unfortunately a big news item, the simultaneous postponement of the CHOGM meeting and the calling of the federal election, effectively wiped the launch off journalists' radar. Undeterred, Board member Rick Farley made it a joyous occasion. 'What I like most about this initiative is its practical approach – if there is important country, go out and buy it. Get around all the red tape regulations by taking control yourself. It is a very simple concept, but devastatingly effective.'[30] And in his introduction, he paid his respects to the Aboriginal owners of the land, a new approach at the time. This public acknowledgment of the Aboriginal ownership of land made Humann realise that Bush Heritage needed to be more attuned to Aboriginal aspirations and the possible role that Bush Heritage could play. For Humann it was a 'light bulb' moment, suggesting how they might involve the Traditional Owners at Carnarvon, with whom they had had little interaction. Farley also addressed the connection between land conservation and the environmental challenges facing Australia, speaking about Carnarvon as 'a very inspiring example of community initiative – of people who have put their values into action', and saying that individuals and groups could make a difference.[31]

Toyne was delighted with the success of the Carnarvon Station acquisition and the response of the public, particularly through donations, as it eased the Board into believing in a much bigger future. With Carnarvon Station, Bush Heritage defined its position as the first national organisation in Australia to focus on the acquisition and management of private land, and as a serious player in the conservation business. And the positive public response reassured those who had doubted the wisdom of the purchase. Perhaps increasing concern over climate change underlined the urgency to protect bigger landscapes, and, rather than turning people off, the 'risky' purchase attracted more interest in Bush Heritage than ever before. Financial support from current and new donors poured in, and for the first time incoming revenue topped $1 million. In that buoyant atmosphere, restructuring the organisation seemed possible.

Structural change came over several years. Not everybody agreed that change was necessary, but Toyne's desire to prove the credibility of Bush Heritage on the national stage meant that pushing ahead was the only option. Injecting new energy and expertise into the Board, by reworking the constitution and examining the role and tenure of the president, was vital. Toyne proposed a maximum of three terms of three years for the role, arguing that nine years was enough for one person to be at the helm. This replaced a clause providing for life tenure in the position. Toyne's presidency was a time of growth, and the coming of age for the organisation.

Once again, staff were welcomed to Board meetings, with Toyne expecting them to present reports, and ask and answer questions.[32] There were clearly mutual benefits in this form of exchange. Toyne's relaxed style emphasised working cooperatively, and many commented on his skill at defusing tense situations with a light touch and humour.[33] With serious questions and decisions to be made affecting Bush Heritage's long-term viability, Toyne also nurtured that culture of inclusiveness and sense of 'family' which was second nature to Bob Brown. This approach was endorsed by an unexpected visitor from the US.

In May 2001 Kent Wommack, seemingly at the height of his career as Executive Director of the Maine Chapter of The Nature Conservancy (TNC), the largest charity in the US and the largest environmental organisation in the world, spent a three-month sabbatical in the Hobart office of Bush Heritage. It was a surprising move, partly explained by Wommack's desire to help others achieve goals in conservation. Bush Heritage staff and Board members had a unique opportunity to learn from a master at a significant time in Bush Heritage's development. Perhaps the most important aspect of Wommack's mentoring visit was that he inspired confidence that they were on the right track, and could do more.

A valuable mentor

Kent Wommack's entire working life has been spent with TNC. His actions as Director of the Maine Chapter provide insights into the thinking and complexities of the vision that has made TNC the major force for conservation in the world. He had in 1987 represented TNC and other Maine conservation groups in negotiating a deal with a logging company to preserve 50 miles of shoreline on twenty-four lakes.

DIAMOND OCCIDENTAL FOREST, a French company, was keen to log, sell or develop 800 000 acres of the Maine forest, and Wommack and Daniel Christensen, Diamond Occidental's Vice President, spent over thirteen hours 'horse trading' different properties, earning Wommack the 'Iron Pants Award' for staying at the negotiating table longer than anyone else.[34] Twelve years later, in 1999, Wommack completed a massive campaign for TNC. It was the biggest campaign in its history, exceeding the target goal of $35 million by $22 million, and it changed the way TNC viewed conservation.

TNC, through the Maine Chapter, had agreed to partner a logging company in buying a huge tract of forested land. As collaborator, TNC would acquire 30 miles of frontage of the iconic St John River for conservation and recreational purposes. Six weeks before settlement, the deal collapsed, and Wommack, keen to retain the river frontage, persuaded the central board of TNC to make a counter-intuitive move and buy the entire 185 000 acres, including the timber mills. He suggested TNC play a bigger role by embracing the needs of all the stakeholders, including their natural foes, the timber industry. He obtained TNC central board approval, promising to raise $10 million in six weeks to repay the initial loan from the board, which he did. By buying land they did not want and could sell off, they paid the $7 million interest on the loan, and further income came from working the timber mills. Wommack was empowered by the new approach:

> We cut trees, but at half the rate of the previous owner, and because we actively worked (out of legal necessity) with the landscape that included people and businesses, people reacted favourably. Several years later we cut into another deal with a paper company on the verge of bankruptcy. We wanted to buy some

of their land, they couldn't sell it to us because it was collateral for their $50 million loan and at risk of foreclosure. They were preparing to sell the lot, 1100 employees at the mill were going to lose their jobs. So we went to the company holding the loan and bought it, carved off the piece of land we wanted, refinanced the company, saving them millions of dollars in cash flow and saving hundreds of jobs.[35]

The 'holiday' at Bush Heritage was perfect for Wommack; he needed 'time out' after completing the campaign, and Bush Heritage was attempting to scale up its organisation. Wommack recognised 'kindred spirits going through similar experiences. It was fun to be there as this young organisation was reaching for the next level. I had just taken a huge risk that turned out to be wildly successful, so I knew it could be done'.[36] He estimated that, philanthropically, Australia was fifteen to twenty years behind the US, and he 'knew how the story would end'.[37] His assessment of Bush Heritage's potential was positive. On being introduced to the Board at Toyne's second meeting as President, he described how a highly skilled and professional outfit could make an impact, exhorting Bush Heritage to accelerate its activity to save the natural environment before it was too late. Success would only come when everyone on the Board was engaged by the vision, so it was important to find the 'right' people. 'Soft spoken and unassuming', Wommack was inspiration personified.[38] He 'opened the door for us and gave us a series of strategies and showed us what we could do'.[39] For Kate Fitzherbert and Roewen Wishart, on the revamped fundraising team, his advice on fundraising for major gifts was invaluable: 'There's the art of being quiet. Once you have put "the ask", don't say anything more. The next person who speaks loses'. With self-deprecating humour, he added: 'I have sat through some very long, uncomfortable silences!'[40]

Wommack's report on Bush Heritage for TNC set the tone for collaborative and productive years between the two organisations. In 2001 Wommack assessed the threats to the environment in Australia as 'serious and urgent':

> Only a small fraction of Australia's forests and grasslands remain after 230 years of European

Kent Wommack from TNC visiting Gondwana Link, WA, in 2006

Source: Bush Heritage Australia

settlement, yet the country is still among the greatest land clearing nations by volume, currently ranked sixth in the world. And, although the environmental consequences of land clearing are well established, the rate of clearing is in fact quickening, especially in the northern tropical state of Queensland.[41]

He observed that Bush Heritage was Australia's 'only NGO with a national mission of biodiversity conservation ... led by an extremely capable and experienced CEO, a strong and well-connected Board of Directors ... staff are relatively young but highly professional and competent ... all are deeply committed to becoming one of the nation's leading organisations for biodiversity conservation'.[42] He recommended that TNC formalise its relationship with Bush Heritage as a key partner in Australia, and that TNC assist Bush Heritage in private fundraising, and in its quest to have tax policies modified to advance conservation.

Back in 1996, TNC had been invited to investigate the possibility of supporting environmental agencies in Australia by the Australian entrepreneur-turned-philanthropist David Thomas. Thomas had owned the online wine business Cellarmasters, which he sold very profitably. A keen fly fisherman, he holidayed in remote fishing lodges in the US, where he met other fishermen who spoke reverently about TNC, lauding its work in protecting wild regions. The men were mostly Republicans, and Thomas had not associated Republican politics with environmental, green sympathies. Being curious, he asked Max Bourke to investigate the link. Bourke, following his stint as a radio presenter, and years as a public servant advising on environmental matters, was now operating his own farming business along environmental lines, while also working with Thomas on his philanthropic foundation. He visited the US to explore the rationale and workings of TNC, and his report inspired Thomas to offer TNC A$10 000 to send someone to Australia to provide advice to conservationists. Peg Olsen and, later, Olivia Millard were the first TNC people in Australia and provided a strong advisory presence. Olsen encouraged Doug Humann and Karen Alexander to attend the Land Trust Alliance Rally in Colorado in 1999, where they were exposed to a dizzying array of networks and collaborations in the conservation sector, and saw sophisticated templates for strategy and governance in action. It was Olsen who suggested that Wommack embed himself in the Bush Heritage office. Through this connection, the Thomas Foundation would later find its way to Bush Heritage, too.

He recommended that TNC formalise its relationship with Bush Heritage as a 'key partner' in Australia, and that TNC assist Bush Heritage in private fundraising.

Among the Bush Heritage staff, Doug Humann benefited most from Wommack's presence, and the two men, of similar temperament and age, became friends. Wommack's time in Hobart strengthened the connection between the two organisations, and encouraged the flow of US philanthropy into Australia. Following Wommack's visit, Humann made annual trips to the US where he met philanthropists from the Maine Chapter, among them the influential Stephanie Speakman, who hosted several functions that connected Humann with Americans interested in environmental philanthropy. American philanthropy, massaged by Wommack's tact and charm and Humann's description of properties ripe for saving, began to flow in for Bush Heritage purchases, further fuelling activity to secure more land. Later, in 2002, when Wommack returned to Brisbane to establish an office for TNC to support several conservation agencies, Bush Heritage would continue to benefit from his advice and support.

Given that the US is a big continent with its own conservation issues, why did US philanthropists choose to buy into Australian landscapes? Significantly, they were attracted to land purchase in Australia because the land was cheap ('You mean you can buy an acre of land for $1.40?'), and by the leverage obtained through the NRS.[43]

But the answer really lies in TNC's evolution into an international conservation organisation involved with nature conservancy globally. It offered practical and financial support to conservation agencies everywhere if they met certain criteria. Australia, as one of several threatened biodiversity hotspots in the world, was a candidate for their help, and Americans appreciated the language commonality, while stable government implied protection for their investments. Seen against alternative destinations in Third World countries where outcomes were less certain, Australia was a popular option.

One of TNC's first goals was to preserve typical examples of every kind of natural wilderness, a model Bush Heritage was replicating. It is worth noting a certain irony in the determined use of the word 'wilderness' since, even then, it was recognised that no area was actually a wilderness. Manipulation of the environment was an ongoing process, through genetic and climatic interference, and more notably in recent history, by human intervention, since indigenous peoples had continuously shaped the landscape to suit their own requirements. While TNC was keen to protect 'wildernesses' when it could, its mission had changed, and by 2001 it knew that over-exploitation and climate change were threatening the properties they had already bought. The traditional strategy of buying land with high conservation values to manage in perpetuity changed to include a broader sweep of stakeholders and a more sophisticated approach to conservation.

Wommack's efforts in Maine had alerted TNC to what could be achieved for landscapes in which other businesses and industries were involved. Understanding that conservation included built environments and the people who lived and worked in them was a leap forward. Climate change and extreme weather conditions obliterated links between natural reserves, derailing programs by destroying land already saved, and endangering more species than ever before. Wommack and other strategic planners at TNC were seeking different ways to address these issues, and the philosophical and theoretical conversations Wommack had with Toyne and Humann and other Board members would have revealed this approach. Thus Bush Heritage avoided some of the growing pains that TNC had experienced due to the massive scale of its operations during eras when the complexities of conservation were less well understood. The urgency that climate change, burgeoning population growth and rampant consumerism were adding to the conservation debate by 2000 drove Bush Heritage to reconceptualise its own operations at a much earlier stage in its development than TNC had done.

Location, location, location

In 2000 Bush Heritage signed a lease on a new office in Kelly Street, Hobart, after several years in the cramped office space on the corner of Murray and Collins streets. However, the idea that the organisation would eventually leave Hobart and move to Melbourne was still a continual source of discussion and worry for the staff.

THE KELLY STREET OFFICE WAS a prize-winning architectural building cut into the cliff face on Salamanca Place, with five open-plan floors connected by one stairway. The office offered little privacy, while underlining the hierarchical structure that was developing. With a street address on Kelly Street, the Fundraising department and CEO were on the top two floors. Finance was on the next floor down, and Land Management occupied the one below, with egress to Salamanca Place, leaving the basement or 'dungeon' for stores.

Toyne believed relocation to Melbourne was essential. In October 2001 he sounded out the Board. The initial response was not encouraging, with most favouring remaining in Hobart. They believed Bush Heritage would not survive if it moved; that losing staff who did not want to relocate would destabilise and fragment the organisation. The idea of two bases, one in Melbourne and one in Hobart, was also mooted, but again few were in favour, feeling that the spirit driving Bush Heritage would be lost. The fundraisers, however, were more positive, suggesting that proximity to donors was important, and that credibility with institutions, research organisations, contacts and volunteers would be easier in Melbourne. According to Doug Humann, the general consensus was that 2002 was not the right time to move. Toyne wound up the initial discussion by reiterating his conviction that relocation to Melbourne was going to happen.

The October meeting was Karen Alexander's last on the Board. While agreeing with Toyne that a move was inevitable, she noted the associated costs – there would be physical, financial and emotional implications. As discussion continued, the mood changed. Perhaps Toyne's persuasive powers were working overtime – Board members gradually began to accept that this was not a case of 'if', but 'when'. As the meeting progressed, Board support for the move came, with recommendations that Humann and Toyne examine the organisational effectiveness of the dispersed model, and how the move

Staff at the Kelly Street office, 2001. From left: (standing) Julie Marshall, Leanne Bleathman, Janine Dirks, Nathan Males, Stuart Cowell, Doug Humann. Seated: Roewen Wishart, Julie Barnes, Bronwyn Brown, Roy Leeman, Kate Fitzherbert

Source: Bush Heritage Australia

would affect fundraising, advocacy and scientific relationship-building, as well as relationships with other NGOs.

The three years, 2002–05, between making the decision and completing the move were difficult. Humann instigated a 'Retreat' for staff, to foster the camaraderie that was in danger of dissipating, and to include in 'the family' a new layer of employees who had joined Bush Heritage, the first reserve managers at Carnarvon Station Reserve, Steve Heggie and Mel Sheppard. They flew down to join the Tasmanian cohort for three days of discussion and 'bonding'. Judged a success, it was suggested that the Retreat should happen every two years. It was later renamed the staff 'Advance' and then the 'Gathering'. The photograph taken at the 2014 Gathering with over seventy people contrasts sharply with the photograph of the 2002 Retreat, where there were 'not quite enough people to make a cricket team'.[44]

In July 2003 Toyne and the Board began discussions with staff, Bob Brown and Christine Milne, now leader of the Greens in Tasmania, about relocating to Melbourne. Both Brown and Milne were opposed to Bush Heritage leaving Tasmania, and they had the chance to address staff. In some ways the long, painful day, that included one-on-one talks with each member of staff as well as group interviews, was cathartic because everyone expressed an opinion. It ended a tense two-year period that had given false hope that the inevitable might not happen.

The Board tried to accommodate the wishes of staff, with generous offers for relocation, but if they did not come, they could no longer work for Bush Heritage. However, an exception was made for Stuart Cowell when he refused point blank to move. This caused some bitterness, but on being challenged, Humann pointed out that Cowell, who had replaced Blake and was managing the reserves, was seldom in the office, so that it did not matter where he lived.[45] Brown and Milne addressed the staff, and then had a closed session with the Board, where they vented their dismay and fury at the 'destruction' of Bush Heritage. But none of this mattered. Toyne announced that the Board had heard everyone, and the decision was to complete relocation by January 2005.

Most of the Bush Heritage staff did not relocate. Nathan Males was 'philosophically' disappointed that Bush Heritage saw running a national organisation from Tasmania as a drawback. He, ecologist Jim Mulcahy and Greg Blake, who had already left, believed from the beginning of Toyne's presidency that relocation was inevitable and had decided to begin their own organisation, the Tasmanian Land Conservancy (TLC), to maintain and add to Bush Heritage's work in Tasmania. They began their business in the basement of Kelly Street, with Humann's blessing. Males commented that 'Doug was generous and supportive, and tolerant of me running downstairs to manage my second job on his payroll', adding, 'Doug is an interesting character; he is at once highly competitive and amazingly altruistic, a rare combination'.[46]

TLC filled the hole left by Bush Heritage, although its debut in 2001 was not auspicious. Having each put in $50 to incorporate TLC, the three young men were left with $49 to run the business. They also needed fifty members to register as an environmental organisation. It took courage to ask people to chip in for an organisation that did not exist, and it was a year before they acquired $2000 and fifty members, but with no prospect of buying land.

Fortune smiled, however. On the east coast, the owner of property Long Point was keen to see one of his 400 hectare paddocks become a nature reserve. He was the last in his family to farm the land it had owned since 1880. The property was 'wall to wall wading birds and succulent salt marshes, a mudflat covered in samphire that goes bright red, interspersed with little channels and lagoons, a labyrinth of waterways, and down the centre a long relic sand dune with a grassy woodland on it, all surrounded by a huge coastal lagoon'.[47] This magical place had become weedy and degraded, and the farmer sold it for $150 000 – 'A gift, actually', as Males commented. They spent their $2000 on a full-colour brochure explaining their venture and describing the property, and circulated the brochure. However, being completely unknown, they were not hopeful of a positive result.

Within days, the rented post office box was stuffed full with letters containing cheques, a situation that continued

The first staff Retreat at Liffey, Tasmania, April 2002.
Back row, from left: Roy Leeman, Mel Sheppard, Doug Humann, Kalina Koloff, Phil Cullen, Stuart Cowell.
Middle row, from left: Nathan Males, Jos Bentley, Bob Brown, Steve Heggie, Matt Dell.
Front row, from left: Leanne Bleathman, Matt Newman, Natalie Neilson, Kate Fitzherbert
Source: Bush Heritage Australia

for months. They raised the purchase money in three weeks, with enough to buy insurance and a secondhand computer. Contributors indicated disappointment at Bush Heritage's decision to leave, and their desire to see conservation expand in Tasmania. By the time Males left TLC in 2012, it had raised $40 million, and was about to purchase native forests from Gunns Timber after that company collapsed.

TLC's success is significant for the Bush Heritage story. While Bush Heritage was not yet mature enough to formalise the idea of role modelling, and supporting other organisations walking the conservation path, the fact that Humann helped and encouraged the development of TLC exemplifies the approach Bush Heritage would take as it went on to develop partnerships in the future. And because TLC's first property was not a forest property, and its young founders were not imbued with the 'tall tree' ethos of Bush Heritage's founders, TLC was more relaxed about the kinds of landscapes to be saved. They were open to scientist Louise Gilfedder's approach, and followed the conservation issues in the

Staff and families Gathering, Toolangi, Vic, 2014. Too many to name!

Source: Bush Heritage Australia

Midlands with interest. The highway from Launceston to Hobart passes through cleared land with minimal trees. It has been cropped and grazed by farmers for 150 years. But the Midlands, as Gilfedder had discovered, contained patches of remnant forest and native grasses supporting rare and threatened wildlife. Professor Kirkpatrick, with Gilfedder, led a team to develop strategies for conservation in this unlikely area, and TLC support for the project gave it energy. Gilfedder worked on the Midlands Bushweb Project from 1999 to 2003, focusing on the protection and enhancement of native vegetation. The project helped generate a groundswell of conservation awareness, and 213 properties were involved in on-ground works at 520 sites, achieving protection of 6200 hectares of native vegetation and over 70 kilometres of riverbank fencing.

In 2003 Gilfedder and Kirkpatrick, with Hans and Annie Wapstra, authored *Nature of the Midlands*, an unusual and unique collection of essays, articles and oral histories about the natural history of the Tasmanian Midlands.[48] The book was given to landowners to increase their understanding of their land. Between 2002 and 2007 a Land, Water and Wool program was established to integrate biodiversity conservation into sustainable grazing systems, attracting the same farming families. And in 2006 Gilfedder and Kirkpatrick participated in another publication that provided information and management advice for woolgrowers with native pastures.[49]

Through the work of Kirkpatrick, Gilfedder and others, the Midlands became a nationally recognised biodiversity hotspot. The Private Land Conservation Program (the overarching program in Tasmania for private land conservation) instituted a targeted approach to the establishment of voluntary conservation agreements incorporating a range of incentive programs. In 2003 the Non-forest Vegetation Project was established under the

Julian von Bibra is committed to protecting native grasslands on his property in the Midlands of Tasmania.

Photo: Matthew Newton

Private Land Conservation Program to support the protection of rare, vulnerable and endangered non-forest vegetation communities from clearance and conversion to grazing or cropping on all land tenures. The project aimed to achieve this through voluntary conservation covenants with landowners and vegetation management agreeements, with financial and other incentives, and through additional protection measures on crown land. Its message to farmers was that conservation management could exist alongside production management, and biodiversity management could become an economically rewarding enterprise for landowners who were prepared to join the project as partners in a long-term, landscape-scale approach to ensure that biodiversity values remained.

In 2006 Louise Gilfedder won a Churchill Fellowship to travel to the US to investigate partnership models encouraging farmers to incorporate conservation into their farming practices. Her research and contacts facilitated activities focusing on private land conservation in agricultural landscapes, and market-based approaches to sustainable stewardship. Bush Heritage was keen to participate, since by this time it was looking beyond its original model of buying the land, which would never work in Australia because farmers had owned their properties for generations and were not interested in selling.

In 2008 the close relationship between Bush Heritage and the TLC saw them establish the innovative Midlandscapes Plan, to conserve 64 000 hectares of key conservation assets in the area, with the support of Tasmania's Department of Primary Industries and Parks, Water and Environment. The two conservation groups then set up the Midlands Conservation Fund, to support the signing of Conservation Agreements with landholders across an initial 2636 hectares with the goal that ultimately 10 per cent of the Tasmanian Midlands Biodiversity Hotspot would be conserved by 2020.

Julian von Bibra's family has farmed the land for three generations, and he is a staunch advocate for the project. He has committed to grazing and cropping his land more lightly, a change made possible by the stewardship payments he receives in return for protecting threatened natural landscapes. He and his family want a future for their farm, and know they must commit to spending time on conservation activities to protect their grasslands, wetlands, woodlands and the native wildlife, activities which will simultaneously improve arable land productivity. For him it comes naturally:

> The native grasslands represent what most of the Midlands was like pre-European settlement. In winter my father showed us how to make firesticks with bark, and how to light a soft burn, mimicking the way Aboriginal people used to care for the land. Since a large percentage of the little that is left is on properties like ours, we feel a sense of responsibility.[50]

Von Bibra hopes the Midlands Conservation Fund will accrue enough money to allow other farmers to join the project. But the gains are not all about farming, nor about money. For him, 'it's the human side of things that makes it work. Working together with philanthropists has been important. This is not something you can do alone'.[51] Restoring farmland through covenanting and incentives is an innovative approach to conservation and Bush Heritage's support has allowed the organisation to maintain a meaningful and active presence in Tasmania, long after the office relocated to Melbourne. Stuart Cowell oversaw the project until Bush Heritage appointed ecologist Matt Appleby as Midlands Conservation Fund ecologist in 2007, based on his experience in assessing and negotiating conservation covenants with landowners across Tasmania. The Midlands would become one of Bush Heritage's anchor regions in the future, and demonstrate a new approach that added substantially to the conservation results being achieved from the properties that Bush Heritage already owned in Tasmania. Later, Al Dermer who, with his wife Karen Harrland, had begun their Bush Heritage odyssey on a north-west Queensland property in 2004, was appointed to oversee the management of all Bush Heritage's Tasmanian interests.

By 2001 Bush Heritage was looking at recruiting staff in Melbourne and Sydney. Humann knew both Kate Fitzherbert in Melbourne and Roewen Wishart in Sydney and for some time he and Karen Alexander had been hoping to attract them to the organisation. Support from the JT Charitable Trusts for general-purpose capacity-building in the fundraising area made this possible in 2001. Fitzherbert, a trained ornithologist, who had a doctorate in zoology and also enjoyed writing copy, initially joined Bush Heritage as a direct marketer, working with Roewen Wishart who became her boss midway through 2001. Wishart worked from Sydney and built on the professional fundraising program developed by Charlotte Grimshaw, whose year with Bush Heritage had lifted its fundraising profile. Wishart had been working in fundraising and management at Oxfam Australia, so his transition to another NGO was seamless.

Fitzherbert began working from home, and was asked to help find an office to house Bush Heritage. She had visited many unsuitable potential locations before Humann asked her to look at Level 5, 395 Collins Street, on the corner of Queen Street in the heart of the city. Formerly housing the Bank of New Zealand, the building is one of the finest examples of Gothic Revival architecture in Melbourne, its elegant lines and mellowed stone proudly announcing its age and class as a landmark building. Inside, the refurbishment was more homely than efficient, but suitable in style and space for Bush Heritage's low-key ambience.

Fitzherbert could not believe her luck. Office furniture left by the previous tenant, with computers and telephones still connected, was thrown into the deal by the landlord, and the move was accomplished with a City of Melbourne grant, and generous gifts from donors also helped to defray costs. Some years later when the organisation was outgrowing this space, a neighbouring tenant moved out, enabling Bush Heritage to expand simply by unblocking a doorway. The result was several bigger open-plan office areas, connected by a rabbit warren of corridors, meeting rooms and smaller offices – cosy, although not always good for effective communication between departments. This would finally be redressed in November 2014, when the office moved again, to the first floor in the same building, this time with a purpose-built modern office fitout.

Despite initial concerns, the relocation, completed by the beginning of 2005, was soon seen as a brilliant move, and Toyne gained kudos for keeping his resolve. Melbourne produced a large pool of job applicants with impressive and varied skills, who exhibited the same enthusiasm for the ideals Bush Heritage espoused as had staff in Tasmania, so that within six months Bush Heritage had employed enough people to fill the gaps and expand. Although Toyne was initially sceptical about a 'posh' city address, the heritage-listed building suited the Bush Heritage ethos. The location, embedded in the business and legal sector of town, was convenient for developing more effective links with current and potential high-end donors, and with financial organisations.

Back in Tasmania, TLC flourished in the wake of Bush Heritage's departure. Christine Milne never resiled from her belief that Bush Heritage had sold out by moving, but Bob Brown, although his loyalty to the institution he had created was in the balance for a time, accepted the change, later acknowledging it had been in the best interests of the organisation. 'Letting go is important, you just have to move on.'[52] He has remained as Bush Heritage's number one patron, and old wounds had well and truly healed by the time he decided, in 2011, to give his property to Bush Heritage, prompted by his desire to see Oura Oura transformed into a public place in his lifetime. He is delighted with the arrangement that allows him to continue staying there when he wants: 'Bush Heritage has been fantastic about it; I love it here, and it wouldn't have been the same to leave it to any other organisation'.[53]

Meanwhile, the core business of Bush Heritage continued alongside and separate from the rippling undertone created by the relocation effort that took nearly two years to effect. When Bush Heritage moved to Melbourne, it had a suite of new properties, and had made substantial changes to its Board, and to its day-to-day practices. The numbers of new donors were increasing and income from donations was rising. Much of this can be attributed to the energy and enthusiasm that Toyne inspired by his own commitment, and to the dynamic partnership between Toyne and Humann. As Alexander had predicted, their talents were complementary, and they had an instinctive shared understanding of the goals to be achieved.

Humann's special gifts came to the fore at this time. Not every CEO is comfortable in the key role of soliciting major gifts and bequests. But Humann's belief in the objectives of Bush Heritage, his warm and engaging personality, his familiarity with the land itself, and his breadth of personal contacts across government departments, and private and local conservation organisations, meant he was successful in encouraging bequests and in cultivating and soliciting major donations. With Toyne supporting him, Humann spent a large part of his time securing donations and grants ranging from \$20 000 to over \$1 million from individuals in Australia and overseas, from foundations, and from governments.

Humann's special gifts came to the fore at this time. Not every CEO is comfortable in the key role of soliciting major gifts and bequests.

In 2001 climate change and the complex issues of water use in the Murray–Darling River system dominated the news. Those involved in land conservation worried that all available government money would be funnelled into these areas, even though environmental science showed that conserving land with native vegetation actually underpinned effective solutions for both these issues. Advocates promoting environmental philanthropy and private investment in land conservation had been encouraged, however, by Prime Minister John Howard's 1998 initiative to bring business and not-for-profit organisations together. Although originally designed for better outcomes in welfare, the model could include partnerships between corporate bodies and individuals which encouraged a sense of social responsibility towards the environment. Reforming taxation laws to encourage corporate and private investment in the environment was essential, as was a clear demonstration that private land conservation was cost-effective. By 2001 the success stories in private land conservation included the work of TfN Victoria, Bush Heritage, the National Trust (WA), Birdlife Australia and Australian Wildlife Conservancy (AWC).

But many of the land purchases were small in scale. Toyne and Humann agreed that the challenge was building the capacity to develop partnerships with new players to deliver bigger and more successful conservation programs. They considered the revolving fund concept. Bush Heritage could purchase land with high conservation values, place a conservation covenant on the title, and resell the land. The Victorian TfN used this model successfully, but it was difficult for Bush Heritage, since its donors expected ownership of the land in perpetuity, and to engage with the programs and conservation efforts on the properties. Bush Heritage had to find a different model that would work environmentally and financially. Before buying more properties, Toyne and Humann knew they had to improve all aspects of the organisation, and to streamline its processes. Staff in Hobart had been 'drowning' under increased workloads and clumsy, often opaque systems.

Kent Wommack's confidence that Bush Heritage could develop connections to American conservation philanthropy intensified the desire to identify potential land purchases. But TNC required rigorous evaluation criteria, and Toyne was similarly keen to ensure that proper scientific and financial processes were established. However, property acquisition did not wait for everything to be perfect, and, at the end of 2001, several properties were already under consideration, the most important being in Western Australia.

Magic Lake

On our way to the Gondwana Link project we plan a stop at Magic Lake, as on the map it looks like a great destination for lunch. The road takes us through the Great Western Woodland, a green and verdant wonderland of tall trees, rich undergrowth and a profusion of wildly coloured and varied flowering plants, all of which finally falters into nothing as it merges into wheatfields that stretch beyond the horizon. We arrive at the turn-off to Magic Lake. And now, even the wheatfields are no more.

There is only flat land peppered with grey, dead trees, and occasional stands of low brush greenery on thin sticks that wave stiffly in the wind. The place is desolate and dead, no sound except the wind searing across dry whitened plains. We come to a stop at the edge of Magic Lake, a round lake, broad across the middle and framed by a few big trees on the edge of a silvery shore. Venerable old trees by their size. But they are dead, too, and broken branches litter the ground around them.

The wind has whipped waves to a murky brown and we can't see how deep the lake is. I begin to have a bad feeling about Magic Lake, despite its name. Not a great lunch spot, but I will at least walk into the water. I slip off my sandals. The first steps are on silty sand and I venture further. The water reaches my ankles, but sharp angry needles pierce the soles of my feet and I retreat, limping. It is now that I notice on the shoreline twigs encrusted with bright pinpoints that sparkle in the sun. But they are not pretty; they are salt crystals. This land is dead and cruel. Is this what happens when land clearing and intensive agriculture destroy natural soil quality, and salinity takes hold?

Salt-encrusted twigs on the shore of Magic Lake, WA, are vicious on the feet.

Photo: Peter Martin

The Gondwana Link project

In 2002 Bush Heritage bought its first property in the Gondwana Link project in the south-west region of Western Australia, signing on as a partner in the most ambitious conservation project in Australia at the time. But Bush Heritage did more than just sign on, being the first partner to start implementing the process that would, it was hoped, eventually reconnect the Stirling Range National Park to the Fitzgerald River National Park with native vegetation.

THE PROJECT IN TOTAL COVERS a distance of some 1000 kilometres and includes ecosystems ranging from the wet forests of the far south-west to the woodland and mallee regions bordering the Nullabor, with connectivity designed to restore and maintain the ecosystem function and biodiversity of the region.[54] It expresses the approach to conservation that was now driving TNC in the US, and was gaining traction in Bush Heritage through Toyne's leadership and the influence of Stuart Cowell, its conservation manager. But the Gondwana Link project was born of desperation in the face of the WA government's inexplicable actions.

The government's 1980 decision to allow clearing of up to 100 000 hectares of bushland a year had sparked alarm and incredulity among conservationists, farmers, scientists and public servants, who demanded a review. They had thought that the 1950s policy of clearing 1 million acres a year for agriculture had finished forever, since it was recognised as bad policy that caused degraded landscapes, rising salinity and poor economic returns. But the same group that had administered land releases for clearing since 1978 carried out the 1980 review. They scheduled the release of land for clearing in the region north of the Fitzgerald River National Park, although it had already been recognised as an area of scientific and aesthetic importance. Its significance was confirmed when the 1992 Rio Earth Summit listed it in the top twenty-five hotspots in the world for rich biodiversity.

Historically, as mentioned earlier, land clearing had been an Australia-wide government policy. Securing a pastoral lease legally required the removal of all native vegetation to better improve the agricultural potential of the property. Wool prices soared in the 1950s, fuelling land clearance activity, and in Western Australia Colin Cameron, head of the War Service Land Settlement Scheme, rejected 'doing things by halves. There's no use in leaving gaps in your settlement. You must put down one foot firmly at a time'.[55] Land was marked out in square miles, and bulldozers dragging heavy chains and huge wrecking balls knocked down trees and other vegetation within the area, which was then burnt and the land rolled. Newsreels of the day showed images of big trees toppling, billowing black smoke, a troubling match for the triumphalist commentary detailing the taming of the land. Later, more modern machinery cleared everything in one hit, including the small root fibres that held the soil together.

But the removal of native vegetation leaves the soil vulnerable to dryland salinisation and soil erosion, which destroy water quality because of salt and sediment run-off. This is most obvious in Western Australia, where salt content in soil is naturally high, since the land was created from ocean basins and dry sea lakes. Native vegetation keeps the salt below the root line, because root systems absorb the rain. Salt rises to the surface when water penetrates to the salt layer, acting as a conduit, a situation exacerbated by broad land irrigation. When crops are under irrigation, soil remains permanently damp, facilitating the easy diffusion of salt to the surface. Once the crop is taken off nothing stops rainfall percolating down to the salt layer. Salt deep in the soil is endless, and will continue to rise even after it has destroyed all vegetation, including crops, despite the application of more fertiliser and more water. Magic Lake and its environs is an example of complete destruction of a landscape that is now irreversible. The loose soil accelerates erosion, and what begins as a small patch extends with every wind. Sand-drifts blow from affected areas to non-affected areas, so that both the farmer who loses his topsoil and the farmer who finds his good farmland engulfed in sand face financial ruin. Communities suffer, too, since roads become impassable with sand-drift, inhibiting travel and forcing local councils to shift sand. Conservationists in Western Australia were desperate to stop the salinity problem. The key was in the native vegetation, but a new approach was needed. It triggered Angela Sanders and others to envisage linking vegetation systems on a massive scale.

Sanders had earlier conducted the survey and drawn up a management plan for Margaret Robertson before the purchase of Kojonup. She had also assessed other properties for Bush Heritage in the south-west. By 2000 she and Dick Walker were so concerned about the health of the landscape that they met with Susanne and Allan Dennings from the Mallee Fowl Preservation Society to consider solutions outside normal conservation parameters. Dick and Judy Walker, also malleefowl habitat supporters, decided to restore a degraded farm to agricultural use by planting trees and nurturing the natural vegetation. Sanders, aware of efforts in the south-west to link up remnant pieces of vegetation, suggested Walker look at the area between the Peniup Nature Reserve and the Fitzgerald National Park. 'We started talking about linking up that whole area, even though there was only one property for sale, and the project grew from there.'[56] The thinking was ahead of its time, garnering little support. 'Most people thought we were absolutely crazy, it was too big, impossible.'[57] But the project attracted the attention of Keith Bradby.

Bradby worked in the Department of Agriculture as a policy officer in the sustainable rural development program. After working as a beekeeper and then in his

RIGHT **Chereninup Creek Reserve, WA, the first major landcape restoration project tackled by Bush Heritage**

Source: Bush Heritage Australia

FOLLOWING SPREAD **The revegetation at Chereninup Creek Reserve after four years**

Photo: Chinch Gryniewicz

own entrepreneurial direct-seeding business, he had joined the public service, believing that was the best way to improve conservation outcomes. He spent fourteen years in several capacities in government departments while simultaneously operating a private consultancy, Bradby and Mates, that provided advice for government departments involved in agriculture and conservation. While in the Department of Agriculture, he worked with his friend Charlie Nicholson, a senior environmental officer in the Department of Environmental Protection, to thwart legislation that would have opened up the Great Western Woodland for clear-felling. The success of this behind-the-scenes operation brought Bradby into contact with Sanders, and in 2002 he became CEO of the Gondwana Link project. Gondwana Link was groundbreaking because it encompassed conservation on a massive scale, and could only be achieved through collaboration and engagement with a vast number of organisations, landholders and other stakeholders. Most landowners were doubtful about the plan's merit, and Bradby knew that tangible and visible results were the only way to convince them.

Volunteers have played a vital role in Bush Heritage work, including in the revegetation of Chereninup Creek Reserve.
Photo: Amanda Keesing

Why was it suddenly so important to create or protect these huge tracts of natural vegetation? What was wrong with national parks and reserves, which were now regarded, in some cases, as museum pieces? If the complexity of ecosystems that make up the biodiversity on which life depends is to be maintained, the arid to semi-arid areas comprising a large part of the Australian landscape are significant. The soil must be anchored, the water table lowered. Where there is no permanent water, and rainfall is variable and unpredictable, plants will flourish opportunistically, exploiting every few millimetres of rain. The animals, birds and insects dependent on vegetation renewed by rainfall take advantage of these productive moments, often separated by many kilometres, and the ability to move from one feeding and/or breeding opportunity to another can only be maintained through long, wide corridors of vegetation. As changes in climate patterns provoke more severe droughts, more intense wildfires, less rain in some regions and more floods in others, ease of movement becomes unpredictable. Bradby believed connecting the land through restoring the endemic vegetation was 'an insurance policy. There's not much evidence-based science to tell you how it works, but there's a huge amount that says the bigger the area of bush, the more capable it is of withstanding threats'.[58]

TNC in the US provided a significant early donation to Gondwana Link that was allocated to an on-ground project to demonstrate that each funded action was 'worthy in itself', as well as making a contribution to the broader goal. Gaining local support, Gondwana Link attracted additional funding to develop a 'more ecologically and operationally fine-tuned' program.[59] Bush Heritage and Greening Australia were the first Australian organisations to show practical interest in the project, and Bush Heritage's contribution was seminal; it committed to the restoration program – a major planting exercise. Bradby believed the trick was to get going and learn as they went, rather than spending too long devising the grand plan.

Bush Heritage had mulled over the project for two years. It did not fit their selection priorities, since the property was mostly bare barley fields, but it represented the direction Bush Heritage was tentatively pursuing. The threat to the land was clear and the science said it could be saved. Restoring bushland appealed, and it would add to the biodiversity of the region. Their TNC donors were keen to contribute up to $500 000 for the purchase of a Gondwana Link property, and they started with two gifts of $175 000 that Bush Heritage was able to match, and with the possibility of NRS funding they were ready. The project also attracted the attention of Professor Ian Gust and his wife, who for many years had made small monthly donations to Bush Heritage. On a visit to Albany, they fell in love with the area, so that upon reading about the groundbreaking project in the newsletter *Bush Heritage News*, they were keen to see it go ahead. Their five-figure donation helped at a crucial moment. Gust, who was instrumental in the development of vaccines against the hepatitis A virus, would later commit to an even larger project with Bush Heritage.

The funds on the table worked powerfully with the ambition to make a visible impact on the landscape, so Bush Heritage bought Chereninup Creek. It was extremely degraded when Humann first saw it, and he admitted being anxious about the purchase. But one area still retained native vegetation, so they knew what the land could become. They embarked immediately on seed harvesting and planting. Twelve years later the property supports numerous tree species and shrubs, randomly and densely packed at this stage, with some vegetation already over 10 metres high. Not identical to original bush yet, but Bradby's calculations on the success of the project stretch beyond a hundred years, when trees will have thinned out naturally, some growing more strongly, and when there will be appropriate understorey vegetation. Ageing trees will develop habitat hollows, and fallen branches, hollow logs and leaf litter will build habitats at ground level, attracting back birds and small mammals and reptiles that disappeared from the region. Already in 2014, after twelve years honey possums, snakes and echidnas have returned, and numerous birds are collecting nectar from trees and bushes. As Stuart Cowell surveyed the bare paddocks after the first year's planting, and thought about the long-term goals, he summed up the approach: 'You have to imagine the seemingly impossible while doing the possible'.[60]

Changing direction

Keith Bradby believes fervently that 'if you just save what's pristine, you're going to lose it. You have to connect it up to other bushland and get it functioning. It's not big trees generating the biodiversity – it's the stuff below your knees'.[61] The Gondwana Link project debunked the idea that pristine landscape was the only thing worth saving, and Stuart Cowell, coming to Bush Heritage from CSIRO, agreed with Bradby.

COWELL URGED BUSH HERITAGE to develop its philosophy to include these important changes in direction. Board members were already thinking along these lines, and recognised Cowell's potential to spearhead change. He was also supported and mentored by Kent Wommack, so that the transition to a more expansive concept of conservation was much easier than it might have been.

When Cowell took on Greg Blake's role at Bush Heritage in 2000, he continued overseeing all the Bush Heritage reserves, as his predecessor had done. Part-senior ecologist and part-conservation program manager, the job had originally included on-the-ground work such as building fences, dealing with feral plant and animal populations, and making friends with potentially hostile neighbours. After the purchase of the huge Carnarvon Station in 2001, this became impossible and the responsibilities were split, with Jos Bentley taking on the senior ecologist role and Cowell becoming the conservation manager. In this position, he was responsible for all the reserve managers, the ecology staff, making property assessments, and bringing a science-based approach to the problems with which each area was struggling.

Cowell was too young to have been involved in the Franklin River protest, but his roots were in the forests of Tasmania, and he early understood the politics of conservation. He remembered hearing the word 'sustainability' for the first time when working in forestry, and it pricked at his mind, although he realised later that it referred to foresters' jobs. A stint working with Aboriginal people in the Gulf of Carpentaria was followed by involvement in Aboriginal Land Council work and native title in Queensland. Then came a university course on environmental management and time at CSIRO working on land uses, by which time he had a broad understanding of conservation. He recognised that there were many stakeholders in every situation, and not just one model for conservation; he also concluded that it was not necessary to save all the forests in Tasmania: 'Those 100 metre tall trees, yes, they should be saved – no one should be let near them with a chainsaw, but trees that are 40 to 50 metres high, some of those could go'.[62] Concentrating on wet forests meant other ecosystems were ignored, because it was 'pretty hard to tree-sit in a grassland'!

Farley's greatest contribution might have been his encouragement to work with Aboriginal people and to seek ways to help them look after their lands.

When Cowell arrived, Bush Heritage was picking up speed; there were fifteen employees 'and we were all pretty frantic'.[63] He believed time was wasted responding to cold calls about properties. For every property accepted, property managers were assessing forty others, and a case for conservation could be made for almost anywhere in Australia, at either a national, regional or local level. Bush Heritage needed to be more proactive in selecting properties. But this presupposed there was agreement on direction, which involved debate about conservation versus sustainability, about the Yellowstone Park model versus the humanist preservation of the landscape.

For Cowell, conservation included the concept of a sustainability that encompassed the needs of community as well as biological outcomes. His epiphany came from reading Lance H Gunderson's *Barriers and Bridges to the Renewal of Regional Ecosystems and Institutions*, the fourth book in a series (the first was published as early as 1978) in which adaptive management and resilience theory encapsulated the fundamental shift in environmental thinking that began in the 1970s and 1980s, and accelerated in the 1990s.[64] Ecology, rather than biology, became the focus, and systems, rather than individual animals or plant species, were the building blocks. Acquisition of property became the strategy rather than the objective. The Board was changing, and the old NGO mentality that saw conservation as 'locking up' bush reserves was being replaced by a sharper, more professional approach that saw value in a broader approach, and directors were engaging in discussions about other models to achieve their goals.

Before the move to Melbourne, Phillip Toyne had begun developing a new operational mode by making radical changes to the Board membership. In 2002 Rick Farley, Toyne's close friend, inspired his Board colleagues to explore more options and to think outside conventional parameters. For Humann, 'It was Farley's whole listening, thinking approach, his analysis of the Board and its role, and how it had to give support to the CEO. He had a good grasp of governance, and lines of responsibility. That, for me, was the most important thing he did for Bush Heritage'.[65] In retrospect, Farley's greatest contribution might have been his encouragement to work with Aboriginal people and to seek ways to help them look after their lands, although it must be said that some practical engagement with Traditional Owners had begun before either Toyne and Farley arrived. These earlier conversations led in 2003 to Toyne (then President) officiating at the signing of the Memorandum of Understanding (MOU) with the Indigenous Land Corporation (ILC) in Tasmania. The ILC, as the corporate Commonwealth entity established in 1995 to help Aboriginal people acquire and manage land for economic, environmental, social and cultural benefits, is the primary decision-making body, overseeing governance, considering land acquisition and land management proposals for the 25 per cent of Australia that is owned by Aboriginal peoples. To run its programs, it receives an annual payment of at least $45 million from investment returns of the Aboriginal and Torres Strait Islander Land Account.

Signing the MOU was a landmark moment for Bush Heritage, signifying a new direction and its willingness to support Aboriginal groups in conserving their land. From now on, the Bush Heritage story would increasingly be linked with Aboriginal efforts to own their own land in more than name. Toyne and Farley encouraged Cowell as he travelled the country making contact with Traditional Owners, tapping into networks, and into various catchment and management groups with Aboriginal advisers on their boards, to explain how Bush Heritage might help with Aboriginal land aspirations. But it was early days for this kind of liaison; there was both scepticism and suspicion, since conservation organisations could be seen as a force for another kind of dispossession of Aboriginal land.

By 2004 some progress had been made. Bush Heritage staff had identified fifteen ILC properties where there was a likelihood of discovering high biodiversity, and the ILC had agreed to approach the various regions to gauge owners' interest in having Bush Heritage provide consultative help with their management plans. Bush Heritage had also recognised that collaboration with Traditional Owners was essential for the mapping and protection of cultural heritage sites on its own properties, and a project had already begun on the Charles Darwin Reserve in Western Australia. But gaining the trust of Aboriginal landowners was slow.

Cowell spent much of 2003–04 promoting the Bush Heritage–ILC connection within Aboriginal networks, and by 2005 had secured initial agreements with four groups. Agreements were one thing, however, and the hard work began with developing a framework for a good working relationship. Success depended on employing someone of Aboriginal heritage, with a foot in both camps. But this would not occur until 2006.

Meanwhile, Toyne continued to reshape the Board with the 2003 recruitment of Steve Morton, Chief of the Division of Wildlife and Ecology at CSIRO. Morton's presence added scientific credibility, and he was an able chair of the new conservation committee, working closely with Cowell, a former colleague. He instilled a more rigorous view of management, and was at times sharply critical of Bush Heritage's approach, deeming it unscientific. Morton and Toyne had first met in Alice Springs, during Toyne's days as a lawyer and Morton's as a scientist starting out with CSIRO.

Guy Fitzhardinge had become a member of the Board a little earlier, in 2002. His input and advice had already been, and would continue to be, invaluable during the expansion years. As a farmer managing family properties in central west New South Wales, director of a catchment committee in New South Wales, and heavily involved in Landcare, he took a conservationist approach. He understood the sociology and culture of rural communities, maintaining that reserve managers were best recruited from the land, since 'dyed-in-the-wool greenies' had little in common with local communities; he thought it was easier to instil ecological skills in someone who had lived on the land than it was to impart practical skills to someone trained in ecology.

Representatives from the Indigenous Land Corporation with Phillip Toyne (centre back) at Murrayfield, Tasmania for the signing of the MoU with Bush Heritage, 2003

Source: Bush Heritage Australia

The recruitment in 2006 of treasurer David Rickards, Executive Director and Global Head of Research at Macquarie Securities Group; Keith Tuffley, Managing Director and Head of Corporate Finance and Equity Capital Markets, at Goldman Sachs JB Were; and Hutch Rank, Managing Director for Du Pont, provided badly needed high-end financial expertise. Rank supported developing a strong code of health and safety measures, while offering the benefit of his broad business acumen. In 2007 Andrew Myer, a long-time Bush Heritage supporter, experienced in property development, management, film and philanthropy, helped extend the Board's understanding of philanthropy. They all knew that developing confidence in the financial capabilities of the organisation would inspire creative energy in staff, and supporter loyalty.

The path to sustainability was not by exclusion and scientific management from within, but through engagement with communities outside the reserves.

Toyne led the Board through this process of transformation, driven by his desire for Bush Heritage to become the premier national organisation at the cutting edge of conservation. A clever man, always able to respond off the cuff to new ideas, he had a light touch at meetings and Morton believes that Toyne acted intuitively, trusting a board of experts to establish a consistent strategy to support his ambitious vision. Committed to the goal, Toyne 'strode along believing that he could do it, and people were swayed by his infectious enthusiasm'.[66]

At this time three main committees with responsibility for submitting proposals to the Board were created, each chaired by a Board member: one examined the scientific and ecological feasibility of buying a particular property, another examined the financial issues, and the third explored the fundraising potential. They then presented full reports requiring Board approval. The Board was concerned not to lose supporters who wanted to preserve bushland in the traditional way, and there were heated discussions over reformulating the Bush Heritage vision. Toyne did not deal with fine detail, disliking protracted debates, and would curtail discussion with: 'For God's sake, figure that out – just make it happen!'[67] Thus it became Stuart Cowell's role to flesh out the new paradigm.

His first attempt was a program he called 'Beyond the Boundaries'. It exposed philosophical differences between the concept of national parks and the broader approach. Conservation of national parks only worked, Cowell argued, if the threats were static and the tools to manage them were within park boundaries. Tree clearance illustrated this: trees would remain while the initiative came from the reserve management. But other influences and reactions, such as fire, feral animals and weeds, came from outside the reserve and could affect the reserve at any time. The path to sustainability was not by exclusion and scientific management from within, but through engagement with communities outside the reserves. Cowell advocated partnerships with local communities and with reserve neighbours, taking the emphasis off ownership of land, an approach that would open the way to developing partnerships with Aboriginal groups. The concept would broaden to encompass all engagements with other landholders, including Aboriginal communities, as a way of extending Bush Heritage influence around Bush Heritage reserves. But Cowell admits that Beyond the Boundaries was not well received by staff: 'They hated it'.[68] It appeared to contradict what Bush Heritage had been all about in Bob Brown's time. While the Board and Humann supported it, most staff were still working within the framework of the original vision, and their interactions with donors emphasised buying up properties to preserve the native vegetation. Thus, before the Board developed new strategies and communicated these with supporters, it had a more immediate issue to confront: changing the mindset and practice of its own people.

Putting systems in place

This philosophical and directional change was underway when staff were making personal decisions about whether to move from Tasmania to Melbourne, while still maintaining the day-to-day business using unwieldy systems. Delivering reports to a Board now eager for clarity on every issue was still difficult, despite minor improvements, and the mechanisms whereby long-term results could be assessed were not in place.

THE PERSON WHO DID MOST to help staff accept the new vision contained in the Beyond the Boundaries concept was employed as the general manager. And her first job was to manage the relocation from Hobart to Melbourne.

Annette Stewart came to Bush Heritage via a volunteering program. Although weary of her job in finance in Sydney, she did not want, in 2004, to return to the world of information technology where she had begun. It was time to assess her future. In the past she had casually donated to Bush Heritage, and, knowing one of the Board members, decided to volunteer on a weeding program at the Charles Darwin Reserve in Western Australia. During the two months she spent working and managing other volunteers, she met Doug Humann who later arranged to meet her in Sydney. He had seen something in her quiet assurance that might have been overlooked in a conventional interview situation. He suggested she consider the position of general manager at Bush Heritage, which would involve overseeing the move to Melbourne.

Stewart was happy to go to Tasmania. She masterminded the move to Melbourne in a calm manner that eased the dislocation many felt. As a systems-oriented person with experience in big business, moving a small organisation was simple, particularly as she was not emotionally involved with staff who were, in some cases, distraught.[69] Stewart also understood about efficiencies and streamlining activities at the time when the business was exploding in all directions, although no one really wanted to know about these possibilities. The biggest limiting factor for Bush Heritage was always the money available, with staff working on shoestring budgets, and discussion about maximising outcomes provoked hollow laughs.

As general manager for five years, Stewart installed systems to maintain the safety of employees, and to underpin growth that would create efficiencies. But she was disheartened by the failure of her administrative program to keep pace with the conservation work, and wanted to give up. Humann supported her taking time off to think about the bigger picture. After six months away, she realised that fixing things through the Finance and Administrative departments would never work because it was conservation that was the core business. She returned in 2005 with a remit to develop new systems and efficiencies within this area.

Back in 1999, Robert Rosen, then treasurer and Board member, had also been thinking about systems, in his case the financial underpinnings of the organisation. He worried about Bush Heritage's ability to maintain sufficient funds to manage properties into the future, in perpetuity, in fact. If Bush Heritage collapsed, could third parties, such as Parks and Wildlife services, local councils, or community conservation groups manage reserves with the income provided by Bush Heritage investments? Rosen believed in buying properties, but was anxious about their viable longevity. In 1999, 'while very bored in bed with flu I tried to distract myself by thinking about

With staff working in remote places on activities such as controlled burns, the need for a new safety system became critical: Stuart Cowell at Carnarvon Station Reserve, Qld.

Photo: Sajidah Abdhullah

this question'.[70] The result was a paper on the establishment of annualised long-term land management costs, or annualised land expenses (ALEs), for each property, to be invested to provide for costs in perpetuity.[71] He presented the paper to the Board in July 1999.

The concept included all long-term costs, but excluded costs expended in the early years of acquiring a property until those costs had stabilised at a long-term manageable level. Annual amortised capital costs were included; for example, if a fence was estimated to cost $5000 and have a life of ten years, the fence ALE would be $500. He built flexibility into these costs, and called it the volatility index. The costs also included relevant land management overheads, general office overheads, and sufficient overhead absorption to cushion any derailment of Bush Heritage in the future. The goal was to have sufficient investments to provide an estimated yield over inflation to cover the ALEs on all properties held for long-term conservation. ALEs would never actually be spent, so that they would not come into the normal accounting or budgetary processes, thus providing Bush Heritage with the capacity to fund future land management commitments without necessarily prescribing exactly how the organisation would carry them out. He recommended that the Board agree in principle to set aside $800 000 in long-term investments to fund, or notionally fund, future land expenses. In November 2000 the Board meeting adopted the ALEs policy, creating an Endowment Fund as proposed by Rosen, with the reservation that the minimum Reserve Expenses for 2001 be recalculated, since it was felt that the initial calculation was too low.

In 2005 Bush Heritage moved away from the ALEs policy. With many more reserves to manage, the capital amount required was far greater than had been envisaged in 1999–2000, and the annual costs of maintaining Bush Heritage reserves in perpetuity were also already higher. The current (2015) Bush Heritage Endowment Policy was established in 2005, and is based on a requirement to hold three years of reserve management costs in a separate fund to be used for management through a time of organisational difficulties.

Further upgrades to the fund occurred as a result of discussions with staff and Board members in 2014, and in January 2015 the Board approved, in principle, the creation of a new Capital Fund. It was resolved that the fund would be a separately identifiable Bush Heritage fund (and marketed and treated internally as such), but ultimately still be a part of the organisation's consolidated assets. A tender process was run to seek responses from selected investment management companies to take on the role of investment manager for the new Capital Fund, and in 2015 formal governance, rules and investment arrangements were established for the fund, and a charter drafted.

From systems to weeds: Annette Stewart volunteers at Charles Darwin Reserve, WA.

Photo: Jiri Lochman/Lochman Transparencies

That these two developments relating to systems operations at Bush Heritage took hold in the early stage of Toyne's presidency was due to the drive from Humann and the Board to develop a structure to carry Bush Heritage through an ambitious expansion period. They gave Toyne confidence to encourage more purchases. But purchases were still constrained by a small staff carrying a heavy workload, which increased the pressure to develop failsafe health and safety procedures. Now that Bush Heritage had more reserve managers living on remote properties, this issue became a top priority. Health and safety issues were high on the agenda at Board meetings, driven largely by Humann, and later with a powerful ally in Hutch Rank, who was familiar with the health and safety packages that his company DuPont was promoting across industry. In the early days of Bush Heritage, conservation managers and ecologists had criss-crossed the country without much concern, but the remoteness and the hours they travelled worried Humann. He never knew if they had arrived safely at their destinations; mobile phone coverage was non-existent; and personnel were spread too thinly for another employee to be able to help in a crisis. Reliance on neighbours, also usually quite distant, was often the only safety net. This was hardly the way to run a professional business. Infrastructure on the properties was also a concern. The original fire plans had been developed to protect the vegetation, but now they included protection and security for dwellings, sheds, fuel tanks, vehicles, and other machinery. Staff were riding quad bikes, known to have the highest accident rate of any vehicle. They were also, by necessity, using firearms. How could he, Humann, know that they understood how to use them correctly? Were they storing them appropriately, and accounting for their use? How soon, Humann wondered, would anyone know if someone was in trouble?

The safety measures approved by the Board in 2000 were an early iteration of what over time became a comprehensive safety program to protect employees, volunteers, visitors, infrastructure, and Bush Heritage. It eventually included fail-safe provisions for the widespread and remotely connected staff living in some of Australia's harshest environments. But developing a sophisticated system would only begin to happen after the move to Melbourne, when Ray O'Halloran, a health and safety professional, was employed as operations manager to establish safety policies and procedures. Later these were expanded and improved when Geoff Nevard joined the team. Bush Heritage's first comprehensive safety systems, implemented by O'Halloran and Nevard, were operational by the end of 2007.

In the early days of Bush Heritage, conservation managers and ecologists had criss-crossed the country without much concern, but the remoteness and the hours they travelled worried Humann.

Following on from O'Halloran's work on safety procedures, Nevard implemented safety measurement and reporting, as well as hazard and incident reporting. Eve Jani, as the health and safety and risk coordinator, then took the Health, Safety and Environment Management system to a new level. She put in place the trials and use of the Field Safety System that enabled workers to log in and log out on a twenty-four hour, seven days a week basis, a proactive approach tailored specifically to Bush Heritage requirements. Welcomed by Board members, at first it encountered some resistance from field staff, unused to logging in and out when working away from home.

In 2011 professional auditors tested and identified weaknesses in the system, which were addressed. The result, which included an internal review of incident response management for potential incidents such as bushfire, flood, lost persons and workplace incidents, provided a road map that a newly created Senior Management Team then recommended to the Board. New employees were inducted into Bush Heritage safety procedures and, later, volunteers were also required to be familiar with the procedures. Such a comprehensive plan instilled in employees, volunteers and donors a new sense of confidence in the organisation.

PART 3

Once established in Melbourne, a name change gave Bush Heritage Australia new energy, as did reaching out to Aboriginal Traditional Owners who were also striving to conserve their country. Bush Heritage purchased larger tracts of land, and further overhauled its vision and its systems. But the Global Financial Crisis forced a review of its prevailing model of land buying.

Expanding Australia-wide

Carnarvon country

Carnarvon country spills over with names from history begun long before white settlement: Augathella, Babbiloora, Dooloogarah, Warrego. The words curl away into the inky blackness of the road in front. A single light at the shearers' quarters guides us over the last kilometres. Chris and Alison Wilson, the reserve managers, are hosting a barbecue for the Queensland staff team, and we thankfully join the group after the ten-hour drive.

It doesn't look like it on the map, but Carnarvon Station Reserve is possibly Bush Heritage's most remote property. Augathella, the nearest town, is 200 kilometres down the twisting road that becomes impassable black mud in heavy rain.

The morning reveals the homestead valley, and the cluster of buildings on the brow of the hill beyond the shearers' quarters. The grass slopes between are mown flat and smooth. Low hills, some sharply conical, and covered in dense forest, flank the plain that tracks the course of the Channin Creek, now dry, as it meanders back the way we came. We are in a valley deep in the Carnarvon Ranges.

Chris takes us on a tour through lush bluegrass downs, a source of pride for Bush Heritage. The photos of the same area when Bush Heritage took over the property show flat, battered ground with not a blade of grass. Don Royal (a Bush Heritage volunteer who has acted on many occasions as the caretaker of purchased properties prior to the installation of a reserve manager) and Steve Heggie were both here when destocking was taking place and musterers were bringing down rogue cattle with lassos, hard and dangerous work. To Steve, it felt a 'bit like the wild west', the outgoing owners being renowned for shoot-outs with their neighbours. They rode around on motorbikes and horses and wore firearms on their hips – the Carnarvon Ranges having had a long history of cattle duffing, and habits die hard. This is Kenniff country, bushranging country. Back in the 1890s, a policeman and the manager were shot and burnt to ashes as a result of disputes over cattle and horse stealing, and Patrick Kenniff was hanged for the crime. It is still one of Queensland's more notorious and controversial criminal sentences, a northern Ned Kelly drama.[1]

At the Billabong public camping ground a shiny black-headed python about 2 metres long and as thick as my arm slithers lazily across the track, and we disturb two wedge-tailed eagles who fly up in a clatter of wings from the high grass.

Later, Peter and I drive further into the ranges, crossing creeks that trickle water, exploring hidden valleys and brigalow forests, sandy plains and ravines that have cut into the escarpment in tumbles of rocks and twisted fig tree roots, into country of many tracks and disorienting hills, alone in an ancient landscape, safe in the knowledge that we are carrying a Bush Heritage grab bag with a satellite phone and a spot device that will emit our GPS position should we get lost.

We turn back reluctantly into an orange and pink sunset suffusing the sky with a rich glow as we reach the homestead paddock. A thin line of red flame snakes up the side of the triangular-shaped hill behind the homestead, gathering vividness as it bursts into a fireball at the summit, vying for glory with the dying sunset. We see figures blackly outlined against the flames at the foot of the hill, the Wilson kids dancing, as the first cool burn for the season gets underway, and as Chris and others light up the grass along other tracks. It is the first time I have seen a deliberate 'burn' and, always fearful of wildfire, I instinctively distrust firing up the landscape. Chris sees my concern and shows me where the fire will go, over this range and into the next, but no further. 'Won't it just keep burning?' I ask, thinking of Black Saturday. The ranges are dense with fuel here, just as in the Kinglake area in Victoria that fateful summer in 2009. 'It's winter, there's still a dampness in the air and in the grass, you can feel the coolness. We plan it for nights like these. It will just slowly die away,' he assures me.

And he is right. Next day Peter and I drive to Fig Tree Spring. Beside the track is evidence of last night's fire. It wound itself through trees that are blackened but undamaged. Smoke still lingers, and irregular patches where grass has been burnt mark the places that will stall the progress of a big fire.

Cool burns set at the right time are predictable and help limit the impact of future wildfires.
Photo: Emma Burgess

We find the spring, an insignificant hole that would bubble if there were water in abundance, but drought is gripping Central Queensland. A huge callistemon stands guard, although 'stands' is hardly the word. Its massive trunk is too heavy to hold it upright any more and it lies fallen, but still alive, across the spring, its lateral branches pushing vertically again, and broken bits spreading out over the hillside.

But where is the fig tree? I can't imagine anything bigger than the callistemon. We see a slightly smaller fig tree nearby. That must be it. Disappointed, as we were expecting something more dramatic, we are about to leave. The air is heavy; there is a brooding expectancy. We look around.

Hard to know how we could have missed it. Further up the hill, the outspread branches of the fig blot out the sky. If the callistemon is massive, the fig tree has just gone off the scale of adjectives. I am reminded of Enid Blyton's Faraway Tree that always seemed to me the biggest tree imaginable. We scramble up the hill until we are underneath the canopy that spreads an impossibly wide shade, walking around its immense girth, lost for words.

I have read interviews with green activists who lived in tall trees in Tasmania for weeks on end. They admitted to having communed with their trees, and to having heard or felt wisdom being imparted to them.[2] Now I can believe it. We stay silent in the fig tree's presence, feeling the venerableness of its story, its strength in withstanding fire, flood and other unimaginable onslaughts. I try to visualise it as a young sapling, and its later role in the lives of the Aboriginal people of the area, and when that might have been, perhaps 600 years ago? Maybe a thousand? But it is impossible to 'see' another landscape. I send a silent thanks for having been given a moment in its life. I want it to be there for my grandchildren, and I think again about cool burning that might this year save it once more from fire.

The fig tree, 'sentinel of the ages', and ecologist Murray Haseler

Photo: Paul Young

Fences, ferals, fires and floods

Bush Heritage completed the purchase of Carnarvon Station in 2001. At 59 000 hectares, it was the organisation's biggest property to date, with exciting and varied landscapes. Remote and difficult to access, it changed the mindset of the organisation, and all at Bush Heritage began to believe in a bigger future that might encompass landscapes in every state in Australia.

IN A DISTANT PART OF the escarpment that overlooks the country in which Carnarvon Gorge is situated, a huge white limestone pillar towers above everything else. In the shape of a rearing stallion, its front legs punching blue sky in apparent triumph and freeedom, this ancient monolith seems to symbolise the energy and vitality that the property inspired in Board members, employees and donors.

Carnarvon was the first property for which the uncoordinated systems and science that Bush Heritage had gathered over some years came together in a defined management plan. Steve Heggie and partner Mel Sheppard applied over the phone for the positions as joint reserve managers. Then rangers in Kakadu National Park, they were looking for a situation where they could work together, after years of being on different rosters. They were the first reserve managers employed by Bush Heritage, and their practical skills and experience in the field helped create the model for future managers, as Bush Heritage purchases increased and on-site management became the norm.

Bush Heritage had long ago identified its major management issues on reserves as fences, ferals, fires and floods, and Carnarvon had all these in abundance in the first few years. Ferals included all non-native animals, invasive weeds, foreign plants and grasses, with the addition, perhaps, of the slightly more abstract problems of human friends and foes. Given the increasing number of neighbours they were acquiring, it was important to identify who fitted into which category and to understand the interactions, the common issues, the points of difference, and the way that Bush Heritage, as the newcomer, could most usefully fit into the landscape.

Earlier land purchases had required only static conservation, and the land manager, Greg Blake, and ecologists were responsible for carrying out the necessary work. At Brogo, Blake and local ecologist Jackie Miles had restored fences, killed rabbits and eradicated cactus, helped by local volunteers, including the Ipmas, who had sold the property to Bush Heritage. At Goonderoo, there was incentive to take management more seriously, and apply the science. On that property, they confronted, for the first time, the reality of feral animal management, a psychological barrier for several Board members. Even after destocking, feral cattle had to be rounded up, feral cats were rampant, and repairing and building new fences had been a taxing part of the job. Fire management, too, was a source of contention, because the locals used graders to create firebreaks, while the Tasmanian, scientific approach was to slash them. Clearly, climatic conditions dictated different approaches. Goonderoo's previous owner and now neighbour, Hugo Spooner, had explained that they used slashing to carry fire, while Blake maintained that graders would tear up the paddocks. After the first year, Spooner insisted, and Bush Heritage received a good lesson in neighbourly cooperation. As Spooner recalled later: 'It was two different cultures clashing'.[3] Even before pastoralist Guy Fitzhardinge joined the Board, he had urged Bush Heritage to accept local knowledge and to work cooperatively with neighbours, an approach that paved the way for the strategy of cooperation that Stuart Cowell would advocate in the Beyond the Boundaries program.

At Goonderoo, this cooperative approach could have enabled Bush Heritage's participation in the protection of the almost-extinct species of wallaby that Spooner protects on his property. His nature refuge is home to the last wild population of bridled nailtail wallabies or 'flashjacks', once widespread in grassy woodlands from western Victoria to Charters Towers in Queensland, but now confined to Spooner's refuge and two national parks.[4] Spooner had hoped that Bush Heritage's property adjoining the refuge would extend their habitat, but, without cattle grazing to keep the grass down, this was not possible. If lightly grazed or cool-burnt, the grass remains short enough for the wallabies. Spooner chooses the former method, and eventually persuaded Bush Heritage to agist a small head of cattle after fire nearly wiped out the entire property and the house was only saved because locals arrived to protect it.

At Goonderoo, there was incentive to take management more seriously, and apply the science.

While ecologists involved in the management of properties such as Goonderoo recognised that careful and light grazing of cattle helps protect against fire and keeps the grass at a suitable level for other wildlife, this was not so easily accepted by many Bush Heritage donors, some of whom threatened to withdraw large annual donations should Bush Heritage allow cattle, use heavy machinery or firearms, or practise baiting on their properties. Generally, the Bush Heritage strategy of removing grazing had been effective. On Goonderoo it was disastrous for protecting the flashjack and combating the growing population of feral cats. It would take time to reset the strategy, by presenting donors with more flexible approaches. But meanwhile, at Carnarvon, they had to face the possibility of a potential backlash from traditional donors, despite the excitement with which the purchase had been greeted.

Feral animals were a major issue on Carnarvon. With the destocking of the cattle, and after two years of agistment of a small herd to keep the Johnson grass down, brumbies were increasing in number. Damage caused by wildly racing herds was a problem throughout the region; the numbers were so great that the horses were damaging each other as well as land on, and well beyond, Carnarvon Station. Early efforts to control them met public resistance: it was one thing to remove cattle, another to rid your property of 'wild' horses, always an emotive issue.

Overgrazing by wild horses on one side of the fence destroyed habitat for small native animals and plants: Carnarvon Station Reserve, Qld.

Photo: Steve Heggie and Mel Sheppard

The only solution was to cull by shooting. But this was not acceptable to some Board members, nor to many donors, and debates about the situation ensued, before it was resolved with the support of Queensland Parks and the RSPCA. Queensland Parks approached Bush Heritage to be a partner in reaching a solution for this regional problem. Humann and the Board were acutely aware of the potential for, and consequences of, negative publicity, and Bush Heritage's media strategist Alexandra de Blas developed a comprehensive media plan, with instructions for ground staff on how to deal with the public response. And with Queensland Parks and the RSPCA leading a regional approach to the humane and controlled cull, the positive news coverage helped explain the situation to the public. Bush Heritage was seen to be acting in an exemplary fashion, and as an effective partner in dealing with the problem.

With destocking accomplished, the next task was to remove artificial water points. With troughs and dams removed, feral animals were less likely to break down fences to reach water in times of drought, and the natural water flow could again nourish the land. Removing water points would be more urgent in drier regions where goats and camels pose enormous issues. At Carnarvon, they fenced off the many natural springs, some of which serve as sources for the Murray–Darling river system, with barriers designed to keep out the ferals, particularly the pigs, until they were eradicated, but which allowed native animals access to the water. Volunteers contributed many hundreds of hours to this work.

For urban Australians, the extent and complexity of the problems posed by feral animals in remote Australia are hard to imagine, although awareness of the damage to agricultural land caused by rabbits, and the more recent

Long-time volunteers Peter and Margaret Calder fencing Orange Tree Spring at Carnarvon Station Reserve, Qld

Photo: Wayne Lawler/Ecopix

rapid spread of cane toads, highlight the issues. But given the huge numbers of feral animals (for instance, scientists estimate that currently there are between 15 and 23 million wild cats on Australia's continent and surrounding islands), the problem is not how they can be eradicated (that is now an impossibility) but how to manage the numbers to accommodate the least possible damage to native vegetation and native animal species.[5] And here there are two sets of problems that reserve managers and conservationists face. With dingoes removed from many landscapes following baiting, and thus no top-order predators to limit the number of wild horses, donkeys, camels, goats or pigs, the only methods to keep feral animal numbers down are shooting, trapping and more baiting. These issues have to be managed in conjunction with neighbouring farmers, and the most successful baiting programs have involved broad communities and volunteers.

The first priority for Heggie and Sheppard at Carnarvon, and always for any managers on new properties, was the maintenance of existing infrastructure, including fencing, with the aim of repairing, replacing and equipping for future use. Again, volunteers happy to work in remote places like Carnarvon were invaluable. Heggie and Sheppard, used to public service protocol, were delighted by how quickly their requests for various items were met. It seemed as though the organisation existed just for them.[6] They bought new equipment and tools, cleaned the homestead area, furnished the houses, starting from scratch. But they did not think to request a new water supply.

Water came from 20 kilometres away, from a small spring damaged by pigs and horses, and was gravity-fed into quarter-inch rural poly pipe slung through the trees, down creek lines, and sometimes buried, a system that was 'pretty much perished. Every second day we would discover our water had run out or the tank was empty, so we would walk or quadbike or drive, looking for the latest leak, to repair it, almost a daily occurrence for the twenty months we were there'.[7] During the time of the second manager, Mick Blackman, donors on an organised trip were horrified and fully funded a new waterline from a closer well.

Heggie and Sheppard discovered a wonderland of mostly native flora and fauna, as well as a rich Aboriginal culture of rock art and artefacts in hidden valleys and gorges. The region was home for Grahame Walsh, an amateur archaeologist whose later work in discovering and theorising about the unique rock art known as the Bradshaw or Gwion Gwion paintings in the Kimberley region brought him enthusiastic followers, fame and controversy. From 1977 until his death in 2007, Walsh and his team recorded over 1.5 million rock art images and 1500 new rock sites. Before this he had visited Carnarvon Station several times, and his early discoveries of rock art, burial sites and factory sites for spearheads and cutting implements in the Carnarvon Ranges no doubt encouraged him in his life's work on the Bradshaw paintings, and provided Bush Heritage with a trove of information on the rich cultural history of the property.

There was wildlife in abundance, despite the degraded vegetation: yellow-bellied gliders, red-necked wallabies, black wallaroos and huge bat populations. Heggie and Sheppard noted the problems and drew up plans for future management. Working with a local ecologist and Bush Heritage ecologist Phil Cullen, from Tasmania, they established the first five-year management plan for a landscape-scale property. They mapped the property

'Stairways' on fences allow native animals access to the water but stop feral herbivores and pigs, Carnarvon Station Reserve, Qld.

Photo: Bob Crombie

from different perspectives, including the varying geological bases, soil types, different vegetation, wildlife, water sources, prevalence and position of burrows, nests and wallows of feral animals, and the exact locations and types of weeds. They started with a base plan and set priorities, because there was never enough money or staff to do everything.

Bush Heritage people had helped their neighbours fight fires and, when it was their turn, neighbours and local volunteers arrived to help them.

Not much rain fell during their time there, and the Channin Creek never flowed consistently. But in 2010 the rain came in such torrents that the creeks burst their banks, and water destroyed the roads Bush Heritage had constructed for secure access. The sticky black mud acted like a plug, so that valleys were inundated in a metre of water for a year, leaving the managers marooned. The rush of thick growth after the flood gave fire the chance to ravage that same land the following summer – heartbreaking after all the gains that had been made in nurturing the vegetation. Bush Heritage people had helped their neighbours fight fires and, when it was their turn, neighbours and local volunteers arrived to help them.

Work at Carnarvon took up all Bush Heritage's energy, and proceeded at such a pace that it sucked the air from projects already begun at Goonderoo, the nearby property. Spooner, Goonderoo's former owner and now next-door neighbour, was disappointed but not surprised. Carnarvon was stretching the organisation beyond its capacity. It was an exciting property, iconic, picturesque, and massive in every sense of the word, with problems on a matching scale, and work had begun as soon as Bush Heritage took possession. At Goonderoo, the fire trial sites, established to test cool burning principles, were burnt in the first bushfire to go through Goonderoo, which occurred because firebreaks had been neglected and weeds were growing again. Fences fell into disrepair, as did the vacant homestead, which had housed various groups of enthusiastic volunteers and science groups from the universities working on different aspects of the property. Feral cats, always a problem, increased in numbers because no one was living there permanently. Spooner wrote to Humann suggesting he buy back the property because he felt he could better manage it from an environmental point of view.[8] Bush Heritage declined, but its efforts were directed towards making a success of Carnarvon, and all eyes were upon that project.

The lessons learnt at Carnarvon would translate into scientific management plans for future properties, and recognition that local cooperation was essential to combat or mitigate the damage sustained from natural and unnatural climatic events. But most importantly, in those first years, Carnarvon Station Reserve exemplified what could be achieved by taking bold initiatives. For Doug Humann there was gratification, also, that his staff had nominated him for the Equity Trustees' CEO of the Year Award. When the awards were announced, Humann narrowly lost; oddly, what counted against him was that his was a staff nomination.

The Charles Darwin Reserve

The Carnarvon purchase fired the imagination of Bush Heritage supporters and attracted new donors. An early visitor to Carnarvon was Chris Darwin, great-great-grandson of Charles Darwin, father of the theory of evolution, and he fell in love with the place. Chris Darwin, who lives in Australia, had been left a substantial inheritance that he wanted to donate to 'the greatest need in the world'.

HE SPENT A YEAR RESEARCHING the topic, concluding that inequality of food distribution, loss of clean drinking water and mass extinction of species were the prime problems facing the world. Remembering his ancestor Charles Darwin's words: 'I feel no remorse from having committed any great sin, but have often and often regretted that I have not done more direct good to my fellow creatures', he decided to use his legacy to protect a piece of land with high conservation value.[9] Chris Darwin trawled through the biodiversity hotspots in the world, discovering that the southern part of Western Australia supported a huge endemic diversity of plants and animals. When he met Dr Barry Traill, a zoologist and conservationist, they discussed the best value for money in conservation.[10] Traill encouraged him to talk to Bush Heritage and to think about land in the interior. Hence, Darwin's visit to Carnarvon Station Reserve.

Was Bush Heritage interested in buying something that he could support? 'Of course they said "yes". It was music to their ears! Phil Cullen, Jos Bentley and Doug Humann and I went on a wonderful trip; they were very inclusive, and we looked at lots of properties, a really good education for me. Doug is a very impressive guy. I respected him enormously.'[11]

Darwin wanted the property to have significant conservation value, but also yearned for a 'beautiful' property, with an aesthetically pleasing landscape, like Carnarvon Station Reserve. As Darwin had pinpointed the south-west of Western Australia as the region of most interest to him, Cullen approached Keith Bradby for help.

Bradby rang his friend and conservationist colleague, Charlie Nicholson, who, by chance, had just heard that White Wells Station, about four hours north-east of Perth, was for sale. Nicholson knew the property well, having written an extensive and detailed report on it several years before. The comprehensive report traced the geological profile of the land, the recorded rainfall patterns, the history of the Badimaya people whose traditional lands included this area, its days of white settlement, including leasehold changes and different property use throughout its history, extensive descriptions on the flora and fauna of the area, and the prevalence and intensity of the many fire events.

At 68 615 hectares, it would be Bush Heritage's biggest purchase to date. Its size had fluctuated with the many changes to the leasehold over the time of white settlement, and it had been variously a cattle property, sheep station, wheat farm and tourist venue. Beekeeping had been tried, as had sandalwood plantations; at one stage it had been a collection point for eucalyptus oil. None of the ventures was entirely successful, so that the land was never extensively cleared. There were still magnificent stands of salmon, York and gimlet gums and *Eucalyptus wubinensis*, heathlands, salt lakes, wildflowers and flowering shrubs that showed themselves to full advantage in the spring. In most places the soil crust was still intact, indicating that the country had not been 'hammered'. But 2003 was the fourth year of a six-year drought and, while Nicholson understood the potential of the land, he could not vouch for its condition when Darwin, his partner Jacqui and the Bush Heritage group arrived. As he was loading his truck in preparation for the camping trip to show the property off, Nicholson's wife Krista gave him a wrapped parcel, telling him to give it to Darwin's partner. It was a notebook journal with drawings and watercolour paintings detailing the history of the firewood banksia, starting from the seedling growing out of a burnt sandy landscape.

As Nicholson had foreseen, Darwin was horrified by his first sight of White Wells Station:

> On paper it looked fabulous. But it was just appalling. I expected to fall in love with it and it was dusty and overgrown with weeds. There's this horrific thing across Australia called an emu-proof fence preventing emus coming from the desert into crop country. In drought their natural instinct is to go through it. It was like a First World War trench with barbed wire everywhere; emus were caught in it, dying and dead, or wandering up and down to get through – really confronting. I said, 'Oh my God! I don't think I can buy it'. It looked awful, rundown, overgrazed and dry – we were in shock.[12]

During the three days they spent on the property, Jacqui read the book, making notes, as they surveyed the disaster areas. Nicholson realised that Darwin wanted a small parcel of land filled with endangered species – an impossibility. He showed Darwin that the ecosystems on White Wells were intact, that evolution was still happening, that it was a perfect place to teach people to respect nature and understand its importance. But Darwin was sceptical; his friends in Britain might support the project 'if it had lovely rocks, a waterfall, a green tree and a blue butterfly'.[13]

After the trip, Cullen advised Nicholson that, even if Darwin did not contribute, Bush Heritage was interested in buying the property. However, several days later, Darwin rang to commit to the project. Nicholson believed, and Darwin later confirmed, that Jacqui's reading of the book had clinched the deal; while Darwin found it difficult, she had understood the cycle of regeneration, and its potential. After the drought broke, and with a management plan that included a group of devoted volunteer 'weeding mafia' from Perth, the property started to flourish, shyly at first, and then more brazenly as the years of care paid off. Bush Heritage

Gimlet trunks glow in the morning sunshine, Charles Darwin Reserve, WA.

Photo: Peter Taylor

changed the name of the reserve, because Charles Darwin's contribution to the theory of evolution could not be ignored. Chris Darwin was delighted, becoming an ambassador for Bush Heritage, and showing groups of interested donors around the property whenever he could. His enthusiasm is infectious: 'It has been fantastic. I took photos, the before and after shots. It was a dust bowl and it is wonderful to see that conservation does work. It changed, and is amazing now. It is the best thing I have ever done in my life'. And, reflecting on the process:

> I worked in advertising for years and consuming was what was supposed to make you happy, but in fact, giving is what makes you happy. Giving is an underrated activity, and the surprise was how much fun it was signing a cheque, also how many wonderful friends I've made. I didn't expect this.[14]

Nicholson was pleased:

> There is more to be said and this is, I think, the history of Bush Heritage itself, how a movement like Bush Heritage comes into being and how people start to see the country once again with almost Aboriginal eyes, and that this is country that is valuable for just being country, that there is this attachment, the need to respect and appreciate the land for what it is.[15]

Chris Darwin and Jacqui Courtney at Charles Darwin Reserve, WA

Photo: Doug Humann

A broader focus

Backed by the Board and undeterred by staff scepticism towards his Beyond the Boundaries concept, Stuart Cowell continued developing the land management vision that would operate alongside Beyond the Boundaries, which was quietly humming along as Bush Heritage explored different kinds of partnerships.

BUT BEYOND THE BOUNDARIES was still seen as risky in its appeal, because it was advocating a philosophy slightly different to the original vision of buying single properties to protect them. The Board was still feeling its way with partnerships, and decided to keep the strategy low key until it had demonstrated some success.

Time was on Cowell's side. Acceptance of change slid quietly along unguarded fissures in the psyche, and by the time he presented the same ideas in a slightly different package, resistance was fading. Staff were beginning to understand that partnerships did not mean compromising Bush Heritage ideals. Cowell's enthusiasm, his ability to distil intellectual scientific theorising into stories, and his openness to ideas all helped. Long after his departure from Bush Heritage he is remembered with affection and respect.

There was intense pressure: Phillip Toyne and Doug Humann set an almost impossible pace in purchasing more properties. Land degradation and destruction of natural vegetation was occurring at a fast pace and everything had to be done quickly to arrest the rapid rate of species extinction. Over 90 species were already listed as extinct, with declines in the health of hundreds of ecological communities around Australia. And while Bush Heritage's mission was never focused on endangered species per se, conserving the landscape provided vital species habitats.

But how to choose what to purchase, what to protect in partnership with other landholders? How best to manage the land? Cowell took the randomness out of decision-making for property purchase, developing a concept that streamlined processes and established base lines, clothing Beyond the Boundaries in a respectability that had hitherto escaped it. He stressed that Bush Heritage needed to choose wisely to avoid overcapitalisation, and to choose with the aim of sustaining the reserves over a long period. It was therefore essential to focus Bush Heritage's efforts in more clearly defined regions, which Cowell called 'anchor' regions. Within each anchor region, there would be one large and significant property, the 'anchor' property, from which conservation work, in all its many forms, would radiate, and influence the wider region. In this way Bush Heritage would be 'anchored' in the landscape, and Beyond the Boundaries would become the implementation strategy through which to deliver the 'Anchors in the Landscape' vision. Having set these two programs in place, the Board went further. Following the publication of *Sustaining our Natural Systems and Biodiversity* from the Prime Minister's Science, Engineering and Innovation Council in 2002, which stated that an additional 22 million hectares or a further 3 per cent of Australia's land area would be needed to achieve the goals of the federal government's National Reserve System (NRS), the Board decided to work towards protecting by 2025 7.7 million hectares, or 1 per cent, of Australia's land surface and most threatened ecosystems – an ambitious project. Board members knew that it was only by protecting very broad swathes of land that the overall health of all the components – soils, freshwater quality and vegetation cover – could be conserved. If successful, the natural habitats of indigenous birds, mammals, reptiles and invertebrates would be restored, so that they could return in sufficient numbers to ensure their survival. The emphasis had gone from 'buy and manage' to 'landscape protection', pointing to the changed mood on the part of the Board.

The vision gave voice to the growing concern in communities that bush and wildlife were at risk. The nineteenth and twentieth centuries' cultural imperative to clear the Australian bush was being replaced by the twenty-first century understanding that ecologically sustainable land management was essential, and that each region depended on the good health of its neighbouring regions. Health could be achieved by establishing strategic conservation corridors, selected for their existing biodiversity. Described as 'connectivity conservation', these corridors were too long and too wide to be managed by one organisation, and would only succeed if actively maintained at the grassroots level by all the stakeholders.[16]

Anchors in the Landscape swung the debate from property acquisition as the objective to one of strategy, with a focus on bigger landscapes that stretched beyond Bush Heritage reserves and its partnership properties, and longer timeframes. Gondwana Link had opened minds to bigger possibilities, and the experience at Carnarvon highlighted the difficulties in protecting single properties that were at the mercy of regional issues. Just because Bush Heritage bought a property did not mean it had been 'saved'. Bush Heritage was dependent on its neighbours to manage fire, to control feral animals and to take action on invasive weeds. The Anchors in the Landscape concept forced Bush Heritage to choose regions on which to focus for land acquisition and partnerships. The regions were to be selected on sound scientific information and logistical pragmatism, the key attribute being that they contained threatened ecosystems poorly represented in the NRS, so that land Bush Heritage purchased would contribute to the national estate of protected areas.

When the Anchors in the Landscape strategy was launched in 2005, Bush Heritage, on political and philosophical grounds, made clear that, when looking to acquire properties, it would first seek to discuss with Aboriginal Traditional Owners any interest they might have in acquiring the land. Under the landscape approach they would, if invited, offer an opinion, help with purchasing arrangements and management plans, offer training in conservation and ranger activities, and enter into partnerships. Toyne, Humann, Cowell and Board member Steve Morton had all had substantial contact with Aboriginal communities, and were keen for further involvement, but the initiative had to come from the people themselves. Beyond the Boundaries had helped Bush Heritage move away from the traditional conservation model of buying property, giving it the opportunity to develop a more nuanced partnership role that went beyond the traditional position of engagement with Aboriginal people through training and employment as rangers. Cowell's new framework, Anchors in the Landscape, incorporating Beyond the Boundaries, finally quashed scepticism towards the initial project.

By 2005 Bush Heritage had initiated partnerships with several Aboriginal groups. While not everyone agreed, the value in combining science with traditional methods to heal the landscape was well recognised. Aboriginal groups entering into partnerships wanted to learn from Western science to improve outcomes, but they wanted to retain control of their projects and their land. Bush Heritage had already experienced the difficulties of engaging meaningfully with Traditional Owners at Carnarvon. Traditional Owners had initially viewed the purchase of the property with distrust, and the Board understood that moving slowly, without assumptions, was fundamental to developing good relations with Aboriginal neighbours and partners.

The Board decided to work towards protecting by 2025 7.7 million hectares, or 1 per cent, of Australia's land surface

Successive Australian governments were aware that the first Australians had managed their lands successfully over thousands of years, and the Howard Liberal government in 1997 gave practical expression to this understanding when it instigated a new environmental partnership that the NRS program would support. In order to build NRS capacity to protect representative examples of the diverse range of ecosystems, it began entering into agreements with Aboriginal and Torres Strait Islander Traditional Owners who wanted to re-establish effective land management and take custodial responsibility for maintaining the cultural diversity and conservation of their lands and seas. Thus the Indigenous Protected Areas (IPA) program received significant support and IPAs were added as another type of protected area to the established national parks and regional nature reserves. In 1997 the people of Nepabunna Aboriginal community offered 580 square kilometres of rugged limestone hills, siltstone flats, springs and waterholes between the Flinders Ranges and Gammon Ranges national parks to trial this new type of partnership. The land was held by the SA Aboriginal Lands Trust, and in 1998, with Nepabunna community members employed as rangers to manage their country for conservation, the first IPA, the Nantawarrina Indigenous Protected Area, was declared. The declaration was a formal agreement between the government and the Traditional Owners who had initiated it.

The means by which Aboriginal people managed their land has been the subject of much research and conjecture. Bruce Chatwin, in *The Songlines* (1988), highlights the way song gives life to the land, exploring in a semi-fictional way the close relationship between Aboriginal people and their land.[17] Earlier, in 1980, John Bradley began his forty-year journey of discovery with the Yanyuwa families. In *Singing Saltwater Country*, he draws a detailed picture of the subtle and complex relationships the Aboriginal people had, and still have, with their country, where both ancestors and the 'Dreamings' actively take care of the land, maintaining the essential balance that protects and nourishes the environment, in close connection with the weather conditions and their impact on the vegetation, and taking into account the reproductive and feeding habits of the animals in the region.[18] His book distils the years of cooperative engagement with the Aboriginal group who wanted to impart the wisdom of past journeys over their lands before it was forgotten. *Singing Saltwater Country* also helps others to understand the deep physical and emotional connection between Aboriginal people and their land.

The Mabo decision in June 1992 was the first time that the High Court recognised that indigenous people's traditional rights to land was a recognisable, enforceable property right in Australian law.[19] Included in the decision, as part of its reasoning, was the rejection of the concept of 'terra nullius', that the land belonged to no one.[20] This had been the justification for white settlement without compensation since 1788. The accepted view of Aboriginal people as being completely nomadic, hunting and gathering across the land and tossing firesticks in a haphazard way as they went, was now too glib. Indeed, research by both historian Bill Gammage and later, Bruce

WHITE BULL
TAVERN

Pascoe, an Aboriginal writer, presented persuasively in their respective books, *The Biggest Estate on Earth*, and *Dark Emu Black Seed*, points to centuries of care of the land by Aboriginal people who harvested seeds and yams, lived in simple dwellings constructed of wood and mud plaster, built water catchments and sophisticated fish traps, and manipulated the landscape for aesthetic purposes, ease of hunting, and to control fires.[21] Gammage and Pascoe rely on the writings of the early explorers, landscape painters, and settlers who travelled the country soon after 1788, who described the soft loamy soil that 'perpetually reminded [the observer] of a gentleman's park and grounds', with well-grown trees systematically spaced in the landscape, grassy meadows that stretched to the horizon, and forested areas with no bushy understorey that opened out to cleared glades, evidence of landscapes that were carefully tended, across the country, in every state, and covering vast tracts of land.[22] These observers believed they were witnessing a concerted and comprehensive manipulation of the country. Gammage's and Pascoe's conclusions, again drawn from first-hand accounts of those who were there, are that Aboriginal people managed this with a sophisticated and technically complex regime of fire, synchronising burning events with those of neighbouring groups, to maintain appropriate landscapes across the continent. 'What fire to use and when varied across Australia, but the purpose was … to associate water, grass and forest, providing habitats and making the clean, beautiful landscapes dear to Aboriginal feeling.'[23] The explorers and settlers admired what they saw while systematically destroying the evidence, and degrading the soil to a hard-packed, dry consistency through grazing their herds. Later settlers' accounts indicate changes: forests were denser, and bush undergrowth impenetrable. Streams and rivers no longer flowed easily, and the 'park-like' landscapes had disappeared. Gammage and Pascoe attribute this change to the absence of continuous and careful firing of the land, as white settlements spread, establishing pastureland and marginalising Aboriginal people. Their evidence suggests that a partnership of ecologists and conservationists with Aboriginal groups who retain this knowledge and the desire to care for their country will make the best contribution to the ecological health of landscapes in Australia.

Over the last fifty years, terms such as 'caring for country' and 'working on country' have characterised the Aboriginal approach, and land under Aboriginal management has fared, in terms of health and intact ecology, better than many conservation areas managed by governments. Bush Heritage wanted to establish links between Aboriginal knowledge of country and the scientifically based approach. A person with Aboriginal heritage was essential to develop such partnerships.

Over the last fifty years, terms such as 'caring for country' and 'working on country' have characterised the Aboriginal approach.

In 2006 Doug Humann employed Sarah Eccles, who had studied at La Trobe University and trained in Land Resource Management at TAFE. She had worked with Greening Australia (GA) as a land management facilitator, and in a business resource management role in a Koori business network, where Koori people used their traditional knowledge and skills in a modern context. Although Eccles grew up in Tasmania, she is a Wathaurong person from the Bellarine Peninsula in Victoria. Her grandmother, one of the Stolen Generation, had lived on Framlingham Station, later moving to nearby Warrnambool, and had encouraged Eccles from an early age to identify as Koori, so that she spanned the two cultures comfortably. As 'one of those pale-faced crew' living in Tasmania, she felt the loss of connection to her background, and work with Bush Heritage enabled her

Sarah Eccles (right) joins Bidjara Traditional Owners and custodians at Carnarvon Station Reserve, Qld.

Photo: Darren Larcombe

to hear the stories of other Aboriginal people, helping her discover her own identity and Aboriginal roots.[24]

Her position in the first partnership team was co-funded by the Indigenous Land Corporation and Bush Heritage. She and Cowell worked together on Beyond the Boundaries, at that time not yet actively promoted as a key part of the Bush Heritage program. It was new territory, requiring sensitive understanding of the Aboriginal position, and careful explanations as to what Bush Heritage could and could not do.[25] The idea of land rights for Aboriginal groups was not clear-cut, with many having lost connection to the land, with ownership claims disputed, and different opinions as to the way forward.[26] The sharing of knowledge was essential to making Beyond the Boundaries work successfully within the Anchors in the Landscape program. Landscape-scale conservation depended on collaboration with local, state and federal governments as owners and managers of national parks and conservation reserves, and on working relationships with staff and scientists of local catchment management authorities, to enable coordination of management and survey activities with abutting properties, and with other landowners and special interest groups. Anchors in the Landscape, like Beyond the Boundaries, stressed the meaninglessness of boundaries for plants and wildlife so that sharing knowledge, expertise and experience was essential.

Once the connection with Aboriginal groups strengthened, there had to be a common language that incorporated the two knowledge systems.[27] And while Bush Heritage was steeped in traditional conservation principles, it was quickly assimilating the idea that land was something more than just the species and their habitats: for Aboriginal people land underpinned their wellbeing and health, their societal relationships, their very existence – and these concepts broadened the Bush Heritage vision. Medical practitioner Christopher Burgess, whose years of work on the health of remote Aboriginal communities culminated in a doctoral thesis in 2011, coined the phrase 'healthy country, healthy people', reflecting the Parks Victoria strap line 'Healthy Parks Healthy People'. The key tenets in that thesis would further influence Bush Heritage in this field.

In her second week with Bush Heritage, Eccles and Cowell collaborated on a conservation action planning workshop, to develop a blueprint for Bush Heritage engagement with Aboriginal groups. A flexible program, it focused on the idea of 'healthy country', and became the base document for Aboriginal partnerships and for Bush Heritage properties. It helped employees understand that they were *all* involved in the successful development of partnerships, that this work was not the responsibility of just one small group within the organisation. The Anchors in the Landscape program provided the framework to pursue key conservation objectives alongside social, cultural, economic and spiritual outcomes on land owned by Aboriginal people. As their confidence grew, Cowell and Eccles hired other staff to strengthen their partnership work. In 2009 Emma Ignjic took over responsibility for off-reserve Aboriginal partnerships. She was based in Darwin and then later in Cairns for easier access to northern Australia where more Aboriginal partnerships would occur, and Don Bell, an Aboriginal man based in Canberra, worked with on-reserve partnerships. A year later Justin McCaul, also an Aboriginal man, joined the staff as Aboriginal partnerships manager.

It was new territory, requiring sensitive understanding of the Aboriginal position, and careful explanations as to what Bush Heritage could and could not do.

Eccles gained insight into her own Aboriginal heritage while accompanying Traditional Owners who were visiting their country for the first time in thirty years, or who knew the land only from stories. She wishes more non-Aboriginal Australians could experience the palpable sense of continuity and respect for the land that they expressed so clearly.[28] In her first year she was patch-burning savannah lands with a group of Traditional

Owners when they stopped for lunch. The fire was behind them, blowing their way. Concerned that her vehicle was vulnerable, she drove it onto flat rocks in the riverbed, despite assurances that the fire would stop short of it. The fire stopped at the point the Traditional Owners had indicated, confirmation, if she needed it, that they knew their land, and the art of firing it. Properly managed, the mosaic technique draws on a wealth of knowledge – reading weather and wind patterns, knowing how long to leave land to recover between fires, maintaining effective wildlife corridors between burnt patches, and understanding exactly how vegetation will be affected: the distillation of thousands of years of accumulated knowledge and practice, in danger of being lost forever if not captured today. Bush Heritage's involvement in partnerships with Aboriginal people depends on collaboration that combines the Aboriginal approach with the latest scientific research and ecological considerations. And while Aboriginal groups gain scientific knowledge and monitoring capabilities to add to their traditional land management skills, there are reciprocal gains for Bush Heritage in understanding traditional conservation methods that support the protection of thousands more hectares of land than Bush Heritage could ever hope to achieve through its own purchases.

Bush Heritage newsletters began highlighting the importance of Aboriginal partnerships, stimulating donor interest in supporting such work. By 2008 Bush Heritage connections had been made with a number of Aboriginal groups, including Kaanju and Kuku Thypan Traditional Owners, Cape York, Queensland; the Warddeken people in Western Arnhem Land; the Wunambal Gaambera people in the north-west Kimberley region, Western Australia; and the Noongar people who became partners in Gondwana Link, and are involved in cultural heritage assessments across the region. Partnerships took time to nurture; some would not proceed, others would be suspended, and it would be a while before Bush Heritage could feel that tangible progress was being made. An MOU between Bush Heritage and an Aboriginal group was signed only after trust and respect had been established, and this could take several years.

Anchors in the Landscape swings into action

In 2004 the purchase of Ethabuka Station, in Central Queensland on the border with the Northern Territory, brought Bush Heritage its first real encounter with conservation under desert conditions. Apparently flat and sparse, the ripples of low red sandhills striping across the land give the lie to this initial observation.

ALTHOUGH DESERT COUNTRY, it nevertheless supports interestingly diverse vegetation, and from the top of every sandy ridge the vista stretches out, purple-hazed low hills set on red plains dotted with shrubs, grasses and hardy trees as far as the eye can see. It is a haven for desert wildlife, including Australia's largest goanna, the perentie, and an impressive number of other reptile species. The Mulligan and Georgina rivers snaking through the land are mostly dry, except following big rains, but the water run-off from these episodes keeps the Pulchera Waterhole filled during the dry times. A vast expanse of water brimming with shrimps and other aquatic life, the waterhole supports both migratory and resident bird populations, where pelicans, brolgas, corellas and several kinds of ducks and wader birds feed and nest. Designated an important wetland under the Ramsar Convention on Wetlands (signed in 1971 in the Iranian town of Ramsar), which aims to stop the worldwide loss of wetlands, it attracts birdlovers from all over the world. Professor Chris Dickman from Sydney University knows this country well. For many years, his Desert Ecology Research Group has conducted long-term studies of desert ecology, which includes small mammals and reptiles and their population dynamics, on sites at Ethabuka and neighbouring Cravens Peak.

Guy Fitzhardinge and Steve Morton, two recent recruits to the Board, had earlier talked about buying Ethabuka together for conservation purposes, but soon realised that the ideal solution was for Bush Heritage to buy it. When Sarah Eccles visited the property she recognised the strong Aboriginal presence in the land. She would later work with Traditional Owners to identify areas of cultural significance so that they would be treated with appropriate respect and care, and their long-term protection would be assured. One such site, for example, includes a broad expanse of flat land ringed by low rock escarpments, identified by Traditional Owners as a meeting place for large clan gatherings. Paintings on the rock walls of the narrow gorge opening onto the plain confirm the gorge itself as an important ceremonial site.

The acquisition of Ethabuka helped flesh out the new Anchors in the Landscape program on which Stuart Cowell and the Board had been working. Cowell had urged a pragmatic approach in identifying the regions in which Bush Heritage planned to deliver effective conservation outcomes, and where they believed it would be possible to develop collaborative partnerships with other stakeholders in the region, while at the same time paying due regard to the key conservation indicators necessary for an anchor region. The indicators could be the presence of poorly protected or threatened ecosystems, threatened species, or drought and fire refugia among others. Ethabuka gave the Board the impetus for declaring the landscape that stretches from the Gulf of Carpentaria to Lake Eyre as Bush Heritage's first anchor region, encompasssing the arid heartland of Australia, the tropical savannahs, dramatic gorges, and the unspoiled beaches of the southern Gulf of Carpentaria. Although properties in this region have a long history of cattle grazing with impacts most noticeable around water points, the ecosystems are relatively intact because each property is so vast, and relatively isolated from human activity.

The acquisition of Ethabuka helped flesh out the new Anchors in the Landscape program on which Stuart Cowell and the Board had been working.

In addition to the Gulf to Lake Eyre Anchor region, Cowell identified four other anchor regions. Bush Heritage was keen to maximise logistical efficiency and looked to areas where they had already made purchases. The second region included Tarcutta Hills and Brogo, and stretched from the SA border across Victoria, some way in from the NSW eastern coast until it tips into southern Queensland. The third one centred on the Tasmanian Midlands where Louise Gilfedder's work with farming landowners was beginning to show exciting results in restoring native grasslands and woodlands; the fourth anchor region took in Carnarvon and Goonderoo in central southern Queensland and extended as far as Fan Palm Reserve in Far North Queensland. The fifth encompassed a vast area in the south-west corner of Western Australia, covering the coastal region up to Shark Bay and stretching from the Gondwana Link project inland to north of Charles Darwin Reserve, about 300 kilometres north-east of Perth. Once mapped out, it was apparent to donors and supporters that Bush Heritage was taking a new approach to conservation. The graphics showed that individual properties were pieces in a much larger landscape that had to be protected. Cowell envisaged it as a developing program, where new anchor regions could be added in the future if required, but it did not preclude the acquisition of one-off properties outside these regions. Bush Heritage already owned five such properties, and that number would double by 2008. The challenge would be to extend Bush Heritage's presence in these anchor regions with the purchase of further properties, and through developing productive partnerships with other groups involved in conservation in the region. Anchors in the Landscape would pave the way for continuing dialogue with Aboriginal groups in these areas, an initiative close to the hearts of Toyne, Humann and other Board members. It is worth noting here that the anchor regions would later be renamed 'priority' regions.

FOLLOWING SPREAD Ethabuka Station, Qld, a small cluster of buildings in a landscape striped by sand dunes

Photo: Richard Green

Ethabuka was valued at approximately $1.05 million, and its purchase highlighted the developing relationship with the Maine Chapter of The Nature Conservancy (TNC). Doug Humann's annual visits to Maine had been productive, with donors ready to fund purchases, particularly in the arid regions that were only now being considered as potential conservation areas. An additional incentive was the Australian Government's commitment to contribute $2 for every $1 raised from philanthropic sources. The US donors were greatly impressed by this initiative. In 2003 Humann presented the case for buying Ethabuka Station to TNC's Maine board members, with a positive result. One enthusiastic TNC donor wanted to redirect his already-pledged gift of US$300 000 for the purchase. But $50 000 of the gift was required to cover the commitment donors had made to a US conservation project, to ensure its eligibility for the 'Bob Wilson Challenge'. Wilson, an environmental philanthropist who had established a $10 million fund within TNC, matched dollar for dollar other gifts that met his environmental guidelines. Kent Wommack's email to his board colleague, Stephanie Speakman, reveals just how successful Humann's presentation had been: if Maine could cover the $50 000 pledge, the leveraging would be 'quite remarkable'. Investing $50 000 would generate a gift of $300 000 to buy Ethabuka, which would be matched by an 'Australian Challenge' of $300 000, and by Bob Wilson, who would also match it dollar for dollar, 'turning a $50 000 gift to Maine into a $900 000 gift to Australia'.[29]

With these TNC donations and a successful application for NRS funding, Humann was confident they could meet the asking price and begin work on the management plan for Ethabuka. And while it was fine to attract big donors from the US, the purchase was vitally dependent on the support of loyal Bush Heritage donors who responded generously to new propositions. Humann was always impressed by the number of first-time donors who were inspired by a call for help to buy a new property. Bush Heritage still relied on the direct marketing approach that could generate gifts of $10 000 or more, just from a newspaper insert. Every donor was, and still is, important to the success of Bush Heritage's bids for new properties, and for ongoing management costs. One of Humann's roles was to ensure that the Marketing and Fundraising departments maintained the capacity to support new initiatives and keep donors informed.

Bush Heritage had to prove it was a responsible neighbour and, as a new arrival, had to gain the trust of the communities in which it now found itself.

Ethabuka Station, 215 500 hectares on the edge of the Simpson Desert, was intimidating for those with no experience of buying or managing properties of this size. Fitzhardinge, himself a pastoralist, helped negotiate the purchase. He advised sticking to the market value set by graziers, a figure that included the capital works on the property. He would steer the Board through tricky negotiations for some time, commenting that, at that point, Bush Heritage had not grasped the difference between buying a bush property and buying a house in the city.[30] Farmers' souls were bound deep into their land, and there was antagonism in the outback towards 'greenies'. Selling to conservation groups was anathema for most farmers. They considered themselves conservationists, and avoided, if they could, selling to people they did not like or trust. It was important to establish a relationship on shared pastoral experience, and Fitzhardinge warned Bush Heritage against making comments about the adverse effects of grazing or cropping. Pastoralists had properties adjacent to national parks, or knew somebody who did, and were not impressed with their management. Broken fences became the farmers' responsibility and expense, as did the eradication of weeds invading from park land. Bush Heritage had to prove it was a responsible neighbour and, as a new arrival, had to gain the trust of the communities in which it now found itself.

One Australian donor whose imagination was sparked by the Ethabuka purchase was David Thomas of the Thomas Foundation, an Australian philanthropic organisation. He visited Ethabuka soon after Bush Heritage took possession. He was also attracted by the Bush Heritage concept of Beyond the Boundaries, current at that time. Returning from Ethabuka, he attended a Board meeting with Max Bourke, then the chair of his Foundation. Both of them thought that Bush Heritage Board members, while keen to purchase more land, were not setting their sights high enough. By querying its monetary targets, Thomas and Bourke helped the Board take the next financial step up.

Before agreeing to support Bush Heritage for an initial period of three years, Thomas stressed that Board members should contribute financially, contradicting Phillip Toyne, who believed members' time and expertise were sufficient contribution. Thomas asked each member of the Board if they agreed with the Beyond the Boundaries strategy, and if so, were they prepared to contribute to the cause. Members were honest, indicating support for the direction Bush Heritage was taking, but with some saying they would not contribute financially. Thomas also pointed out that if Stuart Cowell were to leave in the next couple of years, he and Bourke could not be confident of the outcomes they envisaged for Bush Heritage. Fortunately, the answers they received from the Board satisfied their critical probing, and Thomas, generous with his initial support, promised more. He also made a three-year contribution of $40 000 per year that enabled Humann to enrol in the Executive Connection program to work with, share and exchange ideas globally with chief executives from a broad range of sectors. Thomas's aim was to extend Humann's executive abilities, believing the CEO should be supported, since he was responsible for the day-to-day direction of the organisation.

Thomas also encouraged the Pew Charitable Trusts to take an interest in Bush Heritage's activities, and Bush Heritage would benefit from this introduction to the wealthy US philanthropic trust. An independent NGO in the US with $5 billion in assets, Pew Charitable Trusts was established in 1948 by the children of Howard Pew, the founder of the Sun Oil Company. The Trusts' aim is to serve the public interest by 'improving public policy, informing the public and stimulating civic life'.[31] Active across a wide range of public policy areas, its environmental vision is to save the natural environment and protect the rich array of life it supports. Emphasis is on the loss of large wilderness ecosystems that contain a great part of the world's remaining diversity, and it operates as an advocate for changing patterns of behaviour, and providing funds for organisations that work to support its objectives. Not interested in owning land, the Trusts' first donation in Australia was $6 million to the Wild Australia Program, to support the Gondwana Link project at the critical time when it was establishing itself. The Pew Trusts would support Bush Heritage initiatives in the arid regions, since its aim in Australia is to secure protection for all areas of high conservation value in the Australian outback, which it defines as an area covering 6 million square kilometres. In 2014, under Barry Traill's leadership, it published the first article in 'The Outback Papers' series, documenting 'one of the last extensive natural regions left on our planet'. Pew Trusts makes an impassioned plea for protection of this region, because 'in an age of diminishing nature, there are few large places left in the world that are still as environmentally intact as the Outback, that offer such a sense of space and allow us to reflect on our fit to the natural world'.[32] It laments species loss, failure of regional communities, the drift by the young towards the cities, the cultural and conservation values diminishing and dying, while 'Australian society little appreciates the extent of such loss: [because] it is gradual, remote, inconspicuous, [and] unknown'.[33]

The $6 million breakfast

Towards the end of 2004, David Thomas and Doug Humann decided that the Thomas Foundation's support for Bush Heritage should be announced in a dramatic setting. Together with Kent Wommack, they began planning for a small party to which TNC would also be invited. The obvious location and timing was Carnarvon Station Reserve, in October, when spring growth was at its best.

THE WEEKEND STAY would include a tour of the property's most iconic sites, and culminate in a Sunday breakfast. The event surpassed all expectations, and would thereafter be known as the '$6 million breakfast'.

Doug Humann, as host, sent invitations to David Thomas and Max Bourke from the Thomas Foundation, to Steve McCormick, the President of TNC, to Kent Wommack and Gerald Miles, also from TNC, and to Gary Tabor from the Wilburforce Foundation, a US organisation devoted to protecting wild lands. An invitation also went to Stephanie Speakman, who, with her husband, had welcomed Humann and provided him with accommodation on his visits to Maine. A member of the Du Pont family, Speakman had also introduced Humann to influential philanthropists, and hosted functions in the US to support Bush Heritage. It was she who had covered the original $50 000 required to facilitate the US donor's gift of $300 000 for Ethabuka. The Bush Heritage contingent included Mick Blackman, the reserve manager; Brian Whelan, a Bush Heritage director representing Phillip Toyne, who was recovering from an operation; Doug Humann; Stuart Cowell; ecologists Phil Cullen and Murray Haseler; and Professor Hugh Possingham from the University of Queensland. Cullen, Haseler and Possingham had gathered much of the ecological, fauna and flora data to support the decision to buy Carnarvon.

The centrepiece of the weekend was the breakfast. After a short walk that took them to a point where they could gaze out over the burgeoning grasslands across the valley floor to forested hillsides and granite escarpments, guests sat down to a delicious cooked breakfast in a grassy area shaded by eucalypts. The sun rising in a cloudless sky offered the possibility of a perfect day. And it was. At the breakfast David Thomas formally advised Bush Heritage of his gift of $3 million over three years, and just as formally invited TNC to match it. McCormick, responding for TNC, undertook to give a response within six weeks.

The breakfast was a game changer for Bush Heritage, a public demonstration of trust from successful and influential business people, and an international NGO. Liberating, too, because it meant that Bush Heritage could actively pursue other land acquisitions without having to chase money, since it enabled leverage of funds from the NRS to support the purchase of Cravens Peak and additional purchases in the Gondwana Link project that were already in the pipeline.

The effect of the $6 million dollars in funding on local philanthropy was profound. It encouraged Australian philanthropists to go public about their gifts, which in turn prompted others to give. Support from powerful NGOs and foundations gave credibility to the growing enterprise, and confidence that its objectives would be achieved, giving rise to more, and bigger, donations. It also sparked the desire to volunteer, to visit and become involved, so that the volunteer program picked up as the financial contributions quickened. Volunteers helped

The $6 million breakfast. Clockwise from left: Phil Cullen, Murray Haseler (obscured), Steve McCormick, Kent Wommack, David Thomas, Stephanie Speakman, Doug Humann, Roewen Wishart (obscured), Stuart Cowell, Peter Thomas (TNC Pacific Program), Max Bourke

Photo: Brian Whelan

speed up the repair and refurbishment of infrastructure on remote properties, and supporters combined their love of the bush with weeding and spraying programs to eradicate invasive weeds, undertook fencing work, and participated in programs to remove feral animals.

TNC took nearly a year to respond to the Thomas Foundation challenge. In 2005 Wommack confirmed TNC's ability to match the Thomas Foundation gift, dollar for dollar, for projects of 'mutual interest to both Bush Heritage and TNC, particularly those activities that advance TNC's conservation priorities in Australia under our [TNC] 2015 [ten-year plan] goal'.[34]

The tardiness of the reply was probably due to the scandal from which TNC had just emerged, and which had dogged McCormick's presidency since 2003. The *Washington Post* broke stories across the US about TNC's covert relations with big business, its gas and oil drilling ventures, and its culture of cronyism, in a series of articles over six days. It painted the picture of an organisation benefiting its own members at the expense of the taxpayer. TNC trustees were upset, justifiably some would say, since the reports ignored the successes and proper dealings that characterised the organisation's long history in conservation. But there was enough truth in the stories to spark a Senate inquiry, and for TNC to examine its own procedures. McCormick spent 2004 and 2005 rebuilding public and governmental trust by facing the charges, admitting faults, and using the scandal to introduce transformative change from the top. He enlisted an outside consultant to advise on governance, banned TNC employee loans, cancelled oil and gas drilling, and suspended the practice of buying from, or selling land to, trustees with connections to TNC as advisers. At the time of the weekend at Carnarvon, McCormick's mind was probably more focused on governance procedures and how to demonstrate transparency in all TNC activities.

Donated properties

In its discussions with pastoralists and others wishing to donate land, Bush Heritage was always conscious of the need for transparency. The land did not always have the required conservation values, and sometimes came with strings attached. Each offer was examined before being rejected with appropriate explanations as to why Bush Heritage could not accept it.

ONE INTERESTING OFFER came before the Board for consideration in 2004. Reedy Creek, over 400 hectares on the central coast of Queensland, would increase the Bush Heritage landholding, but was outside the defined Anchors in the Landscape region, demonstrating the need for flexibility. But more significant in the Board's discussion was uncertainty about managing a partnership with a developer, and concern that environmental aspirations might be compromised.

Board meeting minutes show that the gift was discussed on several occasions before it was accepted. Developer Michael Baevski, brother to Andrew Myer who later became a Board member of Bush Heritage, had bought 628 hectares of land and six beaches from BHP after it ceased sand mining in the area. The Sunrise at 1770 development included a gated community of 170 homes on 120 hectares of coastal strip, an ecotourism resort, an interpretive centre to be run by the local Aboriginal group, and 467 hectares of inland bush, known as Reedy Creek, that was offered as a donation to Bush Heritage. Baevski had undertaken the legal work to cancel the mining leases and to 'sterilise' over 70 kilometres of coastline from any future prospecting or mining, and he was supported by most of the local community. He undertook the construction of pathways linking the different sites within the development while limiting access to areas of ecological significance, creating an environmental precinct protecting the fragile sand dune country where turtles nested. The houses on the estate were built within environmental guidelines established by Baevski, and owners had to abide by strict rules as to the type and quantity of natural vegetation they could or could not remove.

The donation of bushland within the Sunrise project was particularly generous because owners of properties paid an annual 'environmental' levy to fund a Bush

Heritage employee to care for the reserve. Steve Heggie was the first reserve manager, after he left Carnarvon Station Reserve. Matt McLean continued Heggie's work and has built a short interpretive trail through the coastal paperbark forest. The circular walk, a mixture of boardwalks and stepping platforms, has opened up the thick, lush undergrowth, accessing the wetlands, where, on a good day, sightings of the threatened Dunmall's snake, beach stone-curlew and rare grey goshawk delight visitors.

Reedy Creek demonstrates that synergy can exist between developer and conservationist if there is integrity in the relationship, and guidelines as to what will be achieved. As Bush Heritage pursues other partnerships, the agreement hammered out with Baevski stands as an exemplar of accommodation to different agendas and unforeseen benefits. The reserve's accessibility to the general public has been a positive for Bush Heritage. Not everyone can visit the remote properties that Bush Heritage manages. Not everyone will have the sublime experience of taking in the long vistas of subtle desert landscapes; of watching brolgas fly overhead, long legs streaming behind, their haunting calls fading in the still, dry air, as they make for water over the horizon; of nights spent in the desert, gazing into the starry blackness, feeling at one with an ancient land; or of days clambering between granite boulders up green forested ravines to find pockets of wild gardens fed by crystalline streams that trickle their way down to damp valley floors. At Reedy Creek, you can arrive by car, and take the boardwalk across swampy land through a tropical forest. The interpretive boards tell the story of the reserve, and describe what you might discover if you are lucky: rare plants, frogs, colourful birds, insects, butterflies and lizards in the tangled bush. They are there because Reedy Creek is a protected zone. A brief sojourn in a place where natural vegetation is all that is to be seen can open the soul to a different world.

The search was also on to buy a first property in Victoria. This was a strategic decision to find land that

Mel Sheppard and Steve Heggie at Reedy Creek Reserve, Qld

Source: Bush Heritage Australia

Paperbark forest at Reedy Creek Reserve, Qld, remains accessible to visitors, despite flooding.

Photo: Steve Heggie

could be a destination for Victorian supporters and give them a reserve in their own state, while protecting Victoria's threatened grassy woodlands. In 2004 the Judith Eardley Reserve, about 200 kilometres north-west of Melbourne, was launched. The Judith Eardley Save Wildlife Association donated funds for the land in memory of Eardley, a committed wildlife campaigner. Small parcels of land, including the Barnett Block, would be added over the next few years to create a handsome reserve of 1007 hectares called Nardoo Hills Reserve. It protects grassy box and box-ironbark woodlands, and is home to more than 110 bird species, many vegetation communities and plants, including the nationally critically endangered robust greenhood orchid. Given its proximity to Melbourne and the regional cities of Bendigo and Ballarat, it is well served by day visitors and volunteers, who have, among other activities, helped eradicate serious infestations of wheel cactus by the unusual method of 'stabbing' them with herbicide. Nardoo Hills Reserve is the southern end of the south-east Grassy Box Woodlands anchor region, with Tarcutta Hills Reserve, Burrin Burrin and Brogo further north in New South Wales. In 2013 Traditional Owners, the Dja Dja Wurrung, undertook a cultural asset mapping and management project, which revealed important cultural artefacts on Nardoo, such as cutting instruments, clay cooking balls and scar trees (the bark of these trees was removed for ceremonial purposes or to use as canvases for painting). All this adds an exciting overlay to the property. Bush Heritage continues to work closely with the Traditional Owners to manage and protect these sites.

Properties, properties everywhere

The Anchors in the Landscape approach energised the Board. Phillip Toyne's vision of what could be was becoming evident to all at Bush Heritage, and Board members began to consider how to convey this new dynamism to supporters and to the general public.

THE FIRST PART OF THE ANCHORS in the Landscape campaign, designed to raise money to buy properties that would give substance to the strategy, was launched in 2005.

Ethabuka, purchased in 2004, began a flurry of property buying that lasted until 2008, with its near neighbour Cravens Peak purchased in 2005. Eurardy Reserve in Western Australia was also bought in the same year. Cravens Peak, at 233 000 hectares, is roughly the size of the wider Melbourne region. Once again Fitzhardinge advised on the negotiations, and Bush Heritage put together the purchase costs from a combination of US TNC connections, the Thomas Foundation, NRS funding and donations from individual supporters, including a bequest from the estate of Josephine Gay Bell, a Bush Heritage suppporter. Together, Ethabuka and Cravens Peak, separated by a single property, Carlo, have given Bush Heritage an important stake in the Gulf of Carpentaria to Lake Eyre anchor region, with responsibilities to support fragile ecosystems spread over an immense area of desert.

Barry Traill, Director of The Pew Charitable Trusts in Australia, encapsulates in an article in *Australian Geographic* why it is important to have organisations such as Bush Heritage in these desert regions of the outback.[35] Although his observations relate to the Tanami Desert and the Warlpiri people who lived there, they also apply to these properties on the edge of the Simpson Desert.

OPPOSITE Nardoo Hills Reserve, the first Bush Heritage reserve in Victoria

Photo: David Tatnall

FOLLOWING SPREAD Cravens Peak Reserve, Qld, a favourite dune from which to watch the sunset

Photo: Paul Evans

Wreath flower *Lechenaultia macrantha*

Photo: Dale Fuller

The wildlife was once dependent on the Aboriginal people, and thrived in their care. While they were living on country, they burnt the land to provide 'the right habitat mix for different animals, particularly mammals'. When they were no longer there to fire in their traditional way, fires burnt hotter, ravaging much larger areas, and destroying habitats and food sources. Traill also notes the threat to ecosystems in outback regions from introduced, invasive plant and animal species that are running unchecked. The most common are wild cats, rabbits, foxes and cane toads but the list also includes water buffalo, goats, camels, pigs, donkeys, horses, feral cattle, predatory fish in the water systems, and all manner of noxious invasive weeds and grasses that invade where bare ground has been disturbed by grazing, agriculture or fire. No ecosystem will survive, let alone stay healthy, against this onslaught, without regular care and maintenance. With Bush Heritage and Australian Wildlife Conservancy (AWC) actively managing reserves, and with farmers practising good conservation outcomes, there is a chance that vegetation will again be tended, that animal and birdlife will return. Traill believes the more land we have under benevolent conservation management, the better chance Australia has of protecting its unique biodiversity.

Board members had been watching the progress of their partnership in the Gondwana Link project, and were keen to make a greater impact in the South-West Botanical Province anchor region. The positive results from the direct-seeding program on Chereninup Creek Reserve encouraged them to consider buying more land. They joined with Greening Australia in the Great Southern Arc fundraising campaign, tapping into the WA business community through the business and philanthropic connections of Keith Kessell (general manager, Public Affairs, Wesfarmers, and a board member of the WA Symphony Orchestra), and the chair of the campaign, Gavin Bunning (of hardware retailer Bunnings Warehouse). Wesfarmers Ltd had taken over Bunnings, and Wesfarmers CEO and Managing Director Richard Goyder hosted a function for prominent WA people, because he 'liked the idea of getting some of the country back to being the way it was supposed to be'.[36] At the 2005 function Wesfarmers announced a gift of $250 000 a year for three years, ensuring eligibility for naming rights.[37] Bush Heritage subsequently bought five properties near Chereninup. Yarrabee Wesfarmers Reserve, 923 hectares, and Peniup Creek, 2409 hectares, bought jointly with GA, increased Bush Heritage's holdings, encouraging further investment. In 2007 it added Monjebup Reserve, 956 hectares. In 2009 Bush Heritage further increased its stake in Gondwana Link by purchasing Monjebup North, 1107 hectares, and then the 1037 hectare Monjebup Creek in 2014. Monjebup Creek is one of the last remaining substantial areas of remnant bush and has undulating terrain, rocky 'breakaways' and remnant bushland with interesting plant and animal life. A gentle slope overlooking the woodlands offers tantalising potential for the construction of a small building that could be used as a base for groups involved in ecological and scientific research, and for volunteers. These properties completed an important piece of the

Gondwana Link jigsaw in the region between the Stirling Range and the Fitzgerald River national parks. In 2015 the recent ecological restoration of over 400 hectares of cleared land on Monjebup North has already increased connectivity between remnant bush to the south and in the Corackerup Nature Reserve to the north, creating new havens for the vulnerable malleefowl and western whipbird, Carnaby's cockatoo and tammar wallaby. With these purchases Bush Heritage became the most significant partner in the project, thereby gaining confidence to initiate or contribute to similar partnership projects elsewhere.

In 2005 Bush Heritage also had the opportunity to purchase a property north of Geraldton. Eurardy Reserve, 30 070 hectares, abuts Kalbarri National Park, famous for its dramatic redstone gorges hollowed out of the rocks, along which courses the Murchison River that debouches into the sea at Kalbarri. While not possessing the grandeur of the national park, Eurardy Reserve is a key link in the protected area network. With its own special charm in spring, it blazes with colour from a thousand different species of plants. For this reason, owners Margaret and Bruce Quicke had long ago decided not to crop the area. Here nestles the famous wreath flower, flush to the ground, its thick circle of rosettes voluptuously pink and white, in gorgeous contrast to

Eurardy Reserve, WA, is famous for its wildflower display.
Photo: Leanne Hales

Salt-encrusted banks on an arm of Lake Puckridge, Bon Bon Station Reserve, SA

Photo: Glen Norris

its central blue-green stems. Adding to the biodiversity are small stands of magnificent York gums, their angular twisted branch systems stretching from thick, scarred trunks, with ancient hollows providing habitat for birds and unseen small creatures. Years earlier, Eurardy had been set up as a tourism site for wildflower exploration, with several bungalows in the homestead complex and a camping ground. It has its own signpost on the highway, appearing on maps with a camping icon, making it a favourite haven for travellers. Elizabeth George, a long-time Bush Heritage supporter, was instrumental in this development, since it was she who, after discovering the charm of the wildflowers at Eurardy, encouraged the Wildflower Society of Western Australia to undertake a flora survey that documented Eurardy's many wildflowers in a comprehensive field herbarium that is still a reliable reference today. In 2005, when the Quickes decided to sell Eurardy, George was enthusiastic for Bush Heritage to consider it.[38]

In 2006 Boolcoomatta Reserve in South Australia was acquired – a vast property of 63 000 hectares that was part of the traditional lands for the Adnyamathanha people, who were the first to obtain an IPA back in 1998, and whose vision and leadership inspired many other groups to establish IPAs. It is blue saltbush country, with stands of huge eucalypts lining the banks of sandy river courses. In the heart of sheep-farming country 100 kilometres west of Broken Hill and four hours drive from Adelaide, this property has a rich biodiversity, low hills and rocky outcrops, dry desert air, a handsome stone homestead and several historically interesting buildings on the site, including a bluestone shearing shed. In its heyday, Boolcoomatta sheared 120 000 sheep a year. Mentioned in CJ Dennis's poem, 'A Song of Rain', the celebration of the breaking of the drought more than a hundred years ago brings Boolcoomatta together with the welcome sound of first raindrops: 'Patter, patter … Boolcoomatta'. While not in an anchor region, it was a windfall opportunity that came following the SA government's changed decision not to buy the property. Substantial financial support for the purchase came from the Nature Foundation SA, and when combined with funding from the NRS program, made the sale possible under difficult circumstances.

In 2006 Boolcoomatta Reserve in South Australia was acquired – a vast property of 63 000 hectares that was part of the traditional lands for the Adnyamathanha people.

Recalling the Boolcoomatta purchase, Humann acknowledges the help of Fitzhardinge and the federal Minister for the Environment, Ian Campbell. The SA government had already acquired federal funding to buy Boolcoomatta, and was almost at contract exchange stage, when they decided against it. The head of the SA Conservation Department suggested Bush Heritage might buy it. Bush Heritage was keen, it was a wonderful property, and would establish a presence in South Australia. However, there was opposition from local graziers who believed Bush Heritage had received a special deal, while they had been blocked from buying the property themselves. Bush Heritage was the evil presence, and some complained that Campbell was granting funds to a 'green hippy group'. Fitzhardinge chaired a meeting in the woolshed at Boolcoomatta, and a representative of Minister Campbell explained the circumstances of Bush Heritage receiving funding through the NRS. Humann was impressed by the effort that was put into achieving a good outcome, and that the minister supported the NGO against a barrage of complaints from local landholders and conservative members of parliament. It would take time and several exceptional reserve managers to convince the locals that Bush Heritage had something to offer.

A similar situation developed when Bon Bon Station was bought two years later in 2008. The SA government negotiated, was about to exchange contracts, and then reneged. As with the Boolcoomatta purchase, Bush Heritage was approached to acquire the property and, though locked into a sale price it had not negotiated, was keen to proceed. The state government provided one-third of the funds to prove its good faith and, with NRS funding and other philanthropic contributions, including a donation of $100 000 from the Besen Family Foundation for the initial management work, the purchase went ahead. The Stuart Highway bisects Bon Bon Station halfway between Port Augusta and Coober Pedy. This is country opened up for grazing by Australia's most famous inland explorer, John McDouall Stuart.[39] A tough Scotsman, Stuart took minimal companions and supplies as he travelled north from Adelaide, crossing arid gibber plains, shimmering dried salt lakes, and wide creekbeds in which the occasional waterhole gave hope that the group would not die of thirst. Bon Bon, at 216 700 hectares, supports a variety of ecosystems, and geological and vegetation types. With the acquisition of Bon Bon, Bush Heritage now had a considerable stake in South Australia, although the two properties were separated by 500 kilometres. The possibility of establishing a new Anchors in the Landscape region would soon be floated.

Yourka Reserve, bought in 2007, lies at the northern end of the Queensland Uplands and Brigalow Belt anchor region. The property of 43 500 hectares is in prime North

FOLLOWING SPREAD **Wide, open expanses of saltbush country on Boolcoomatta Reserve, SA**
Photo: Boris Hlavica

Morning sun streams through riverside trees, Yourka Reserve, Qld. From left: Paul Foreman, Alexandra de Blas, Ian Landon-Smith, Doug Humann

Photo: Wayne Lawler/Ecopix

Queensland cattle country next to the World Heritage Wet Tropics area, within a designated Australian biodiversity hotspot, because of the array of endemic flora and fauna, notable for its astonishing nature and variety. Rocky headlands and escarpments protect hidden waterways and wet eucalypt forests, home to the rare and endangered red goshawk, of which only 700 pairs remain in the world. Yourka's purchase was made possible by NRS funding and a $1 million donation by retired Sydney businessman and philanthropist, Ian Landon-Smith. Landon-Smith, visiting the property, accompanied by ABC Television's *7.30 Report*, commented that buying the property was one of the most profound experiences of his life: '[As] I walked along this creek bed … with … this terrific dawn chorus of birds, quite suddenly I was overcome by the emotion of it all and the realisation that I'd been able to make a contribution to Bush Heritage to … help them acquire this magnificent property … a tear rolled down my cheek that morning'.[40]

Another significant project for Bush Heritage that began to take shape at much the same time as Yourka (2007) actually had its origins ten years earlier, with the death of Dr Peter Barrer. Barrer was a respected environmental consultant with a 'contagious enthusiasm for conservation of the natural environment'.[41] He lived and worked in the Canberra environs, had been an employee of the NSW National Parks and Wildlife Service (NPWS), and, while conducting surveys along the lower Molonglo River corridor, and working with the Stoney Creek Greenways Landcare Committee, had identified key wildlife corridors and other potential connectivity issues. Barrer's fascination with land corridor connectivity encouraged others to think beyond the scope of individual projects to a far wider area of inclusion. He recommended that the Mt Foxlow–Harrisons Peak area of crown land be converted to Yanununbeyan National Park and Yanununbeyan State Conservation Area for conservation purposes. Although this was not accomplished until after his death, it was an early step in what would later be known as the Kosciuszko to Coast project.[42]

When Barrer's sisters, as executors of his will, made a gift in his memory to Bush Heritage, there was sensitivity among the groups with whom he had worked, since in his own will he had bequeathed funds to be used locally and strategically for conserving important connecting lands. Stuart Cowell, working on Beyond the Boundaries, recognised an opportunity to find a property involving some or all of these communities, taking account of Barrer's own work, and scoping out an area that could be the starting point for a Gondwana Link–style project.

Some time in 2006, Cowell began talking with the ACT branch of GA and others working in grassy box woodland conservation: Rainer Rehwinkel, a senior officer for threatened species in the NSW Office of Environment and Heritage, and Ian Pulsford, divisional manager of Conservation Programs and Planning in the Southern Region of the former NSW NPWS, later the Department of Environment and Conservation. The timing was good because the Anchors in the Landscape concept focusing on the idea of connectivity and landscape-scale work was gaining traction within Bush Heritage, and establishing an anchor property in the south-east endangered grassy box woodland region made sense. Once anchored by an initial property, it could swing the other key concept, Beyond the Boundaries, into action.

Although there was initial scepticism, Cowell pressed on with property assessments to find a starting anchor property with grassy systems in which to invest the Barrer funds.

In Pulsford and Rehwinkel, Cowell met like minds. Pulsford, based at Queanbeyan near Canberra, had been working with colleague Graeme Worboys since the mid-1990s on the idea of a connectivity corridor of forest and woodland landscapes in the region as part of a corridor along the Great Dividing Range and Great Escarpment forests of Eastern Australia.[43] He wanted to work with Cowell and others representing community conservation organisations to explore the possibilities, since it was a huge landscape and any property by itself was just a postage stamp. If located in the right area it could, however, be a significant starting point for a much bigger continental-scale landscape corridor project, another key stepping stone in a corridor potentially linking the Australian Alps to the south coast. At a meeting, convened by Bush Heritage and held at Gundaroo, between several key groups (Friends of Grassland, Upper Murrumbidgee Catchment Coordination Committee, GA and NPWS), they identified, in the grassy forest and woodland area between Bredbo and Canberra in the upper Murrumbidgee River Valley, a connectivity corridor where they could focus on reconnecting remnant vegetation. The first step was to source an initial property in that region. Initially named the East Murrumbidgee Conservation Landscapes Project, at the Gundaroo meeting it acquired the more vibrant title Kosciuszko to Coast, or K2C.

Although there was initial scepticism, Cowell pressed on with property assessments to find a starting anchor property with grassy systems in which to invest the Barrer funds. Long-time Bush Heritage supporter Owen Whitaker helped, while Pulsford and Worboys began developing a proposal and a business plan for the NSW government to consider, an even more ambitious project, which would include K2C. This continental-scale corridor project was initially called the Alps to Atherton Corridor Initiative and later renamed the Great Eastern Ranges Initiative (GER). This plan took in the length of the Great Dividing Range and Great Escarpment of eastern Australia, to be divided into sections.[44] Pulsford believed Cowell's Beyond the Boundaries program, coupled with an initial land purchase, could be the model for forming community-led partnerships to deal with reconnecting very large landscapes. It could also provide a focus point for opportunities in which a range of conservation approaches, and learning from 'doing landcare', could inspire and influence other landholders. Having pinpointed Scottsdale as the most likely property, Humann and Cowell went with Pulsford to Sydney to meet with the Minister for the Environment, Bob Debus, to pitch for support.

The meeting was a success, and government support was promised. One condition of the funding for the K2C

Allan Wills had farmed Edgbaston, Qld, all his life and was reluctant to leave.

Photo: Wayne Lawler/Ecopix

partnership and money to commence on-ground conservation activities with landholders was that Bush Heritage host and provide half the salary for a facilitator to work towards the integration of smaller conservation groups, while offering a range of conservation incentives and programs to interested landholders. Bush Heritage agreed, and Lauren Van Dyke was appointed as the first K2C facilitator, a position that she held for five years, encouraging others to support and join the project.

Scottsdale, located on the Monaro Highway in New South Wales, several kilometres south of Bredbo, was purchased in 2006. An area of 1328 hectares of rolling grassy hills and woodlands with a long Murrumbidgee River frontage, Scottsdale offered opportunities for conservation, revegetation and innovation. David Rickards, a Bush Heritage Board member, made a donation in memory of his mother; the Vincent Fairfax Family Foundation was another generous supporter; and the project also secured NRS funding. Bush Heritage appointed Peter Saunders as manager of Scottsdale, with oversight of the K2C project, and responsibility for making Scottsdale a viable centrepiece to which many different communities could look for ideas for developing their own conservation connectivity. Saunders was an inspired choice for the role, as will be seen.

Scottsdale, and the activities it generated through the Beyond the Boundaries program, was a significant step forward for Bush Heritage, and helped provide a model for establishing other regional landscape-scale conservation partnerships in the GER corridor. Bush Heritage and four other non-government conservation organisations worked with the NSW government to transfer the leadership and governance of the GER Initiative to a partnership between four non-government conservation organisations, with support from the Department of Environment, Climate Change and Water (DECCW). In 2010 under an MOU between the parties, GA, The Nature Conservation Trust, OzGREEN and the National Parks Association took on responsibility for the ongoing success of the project. Bush Heritage did not sign the MOU which related to the whole GER Initiative, preferring to adopt a more cautious approach. At the same time, the title of Beyond the Boundaries was dropped in favour of the Anchors in the Landscape, a concept which now included most of the detail that Beyond the Boundaries had initiated. Recognising that its meagre resources were already stretched, Bush Heritage would concentrate its efforts on restoring Scottsdale, trusting that success in this area would help inspire others to develop the bigger project.

The last major property bought during this expansionist period was Edgbaston in Central Queensland, home of the red-finned blue-eye fish. Eighty-year-old Allan Wills, third-generation farmer on Edgbaston, did not want to sell, but his wife was tired of the hard remote life. Edgbaston, 8074 hectares, lies in marginal farming country, and it had always been so. Parts of the property are like dry marsh land, with patches of groundcover, sparse tussocky grass, and stunted melaleuca, bent and toughened by hot summers and a driving wind that sweeps across the sandy salt-stained plain. The land inclines a little upwards, and beneath the rocky escarpment marking its border with the adjoining property Mitchell grass plains provided the Wills's cattle

Dr Adam Kerezsy worked for five years on Edgbaston, Qld, to stave off the extinction of the red-finned blue-eye.

Photo: Adam Kerezsy

The red-finned blue-eye

Photo: Adam Kerezsy

with better sustenance, although when Bush Heritage bought the property, these, too, were severely degraded. But it is the lower, unpromising paddocks that have attracted the interest of conservation groups since 1990. Water that is a million years old pumps up slowly and regularly from the Great Artesian Basin into shallow desert springs, the perfect environment for two unique species of fish, fourteen unique species of snails and four unique species of water plants. When Peter Unmack, a young amateur natural historian from Melbourne, visited the property in 1990, he realised that one of the fish was unlike any other he had ever seen. Further analysis revealed it to be the red-finned blue-eye, a species of fish found nowhere else in the world.

When Peter Unmack visited the property in 1990, he realised that one of the fish was unlike any other he had ever seen.

Allan Wills had known about these fish all his life, had even kept some in a fish tank in his bedroom as a boy, and although he probably did not graze cattle near the cluster springs because the land was degraded, it is possible that his fondness for the pretty fish kept them safe until they were classified as one of the rarest fish in the world. They are an ecological miracle that has survived in a few small springs on this property for millions of years. This was true of the Edgbaston gobi, too, but the red-finned blue-eye, because of its distinctive colouring, stole the limelight.

Most of the springs were full of gambusia, the small voracious and prolific fish that had been imported from the US in 1925 to eradicate mosquitoes by eating their larvae. Gambusia failed in its task, and because of its ability to produce hundreds of live young, fast maturation rate, flexible behaviour and broad environmental tolerances, has invaded all inland water systems, wreaking havoc on native fish.

After the discovery, Rod Fensham, then working at the Queensland Herbarium, brought his PhD students to Edgbaston to research the distinctive habitat and its inmates. He wanted a conservation group to be involved at Edgbaston. The gambusia was on the point of eradicating the red-finned blue-eye, and a strategy for protecting the fish and other unique species in the shallow pools was essential. However, Wills was a friend, and Fensham could not push him into selling. Wills vacillated for three years; the Queensland government tried to buy the property, and offered substantially more than Bush Heritage would, but Wills, offended by their attitude, declined at the last minute. Fensham thought AWC might buy it, but Bush Heritage was already interested. Fitzhardinge who also knew Wills, finetuned the sale for Bush Heritage, by agreeing that the house would be left as it was, so that the family could return whenever they wanted. Bush Heritage had already reached its quota of pastoral leases in Queensland, but everyone concerned with conservation was keen for the sale to succeed before Wills changed his mind. Allens, the lawyers who worked with Bush Heritage on a pro bono basis, undertook complex legal negotiations with the government, which then changed the provisions of the lease, enabling Bush Heritage to buy Edgbaston.

Today the homestead is dilapidated; pots still hang by the stove, and there are hats on pegs at the door, as though the occupants have just walked out and will be back soon. But they did not return. The workshop is as it has been over three generations: the earthen floor is deep in detritus, old screws, nails, rusty saw blades, and tools, discarded pieces of wood, metal rods, old spray cans and a thousand other useful items used to patch and reassemble and hold old machines and other essential farm equipment together – a testament to the 'make do' temperament, born of necessity in hard times.

Back at the office ...

Despite the high emotion and turmoil associated with the move from Tasmania, Toyne's and Humann's enthusiasm for purchasing property, supported as it was by the recent $6 million breakfast, launched Bush Heritage into a buying spree that lasted from 2005 to 2008.

EACH NEW PURCHASE MEANT that more resources for conservation management were needed, and another reserve manager had to be employed. This of course required more funds, which in turn necessitated an increase in the fundraising and administrative staff.

In 2005 thirteen new staff took up positions in Bush Heritage within a few weeks of each other, so it was not surprising that a certain amount of disorganisation ensued as they settled into their roles. Paul Foreman, a Forestry and Ecology graduate from Melbourne University, remembers the time as both stimulating and confusing. 'There was so much going on, with Bush Heritage in the throes of expanding, and at the same time attempting to deal with processes to support rapid growth.'[45] Having applied for the position of conservation manager, it was months before he learnt he had been successful. Work was stressful and challenging, but exciting, since he could see that Bush Heritage was attempting conservation by working with large tracts of land rather than properties, a relatively new approach that his ecological training told him could make a difference. Even before officially beginning work, he was invited on a trip to Carnarvon and soon after became responsible for effecting the handover of Eurardy, putting together an inventory, and employing the first reserve managers, experience that he later put to use on two other properties, Bon Bon and Boolcoomatta, both in South Australia. He was responsible, too, for finding a property to expand the holdings in the newly designated Anchors in the Landscape Einasleigh Uplands region, in northern Queensland. He found Yourka, presented the case to the Board, which approved the purchase in 2007, and employed Paul and Leanne Hales as reserve managers for the property. Coincidentally, he had originally appointed them as reserve managers for Eurardy in 2005.

Katrina Blake, also appointed at this time, was looking for a role in natural resource management, and thought that she could work her way into that area from the visitations officer position. Even before she started, her name was on a letter sent out to 2500 supporters, inviting them to sign up for a visit to one or other of the Bush Heritage properties. On her first day at work, she discovered that the first scheduled visit to a property was just three weeks away, with more visits in the pipeline. There were fifteen reserves and an oversupply of enthusiastic donors prepared to put their holiday plans in Blake's hands. The schedule was frenetic, with little time to organise trips in advance, and she took groups out without knowing the terrain, the people on the ground, or much about how they were going to get there. After six months, she had accumulated ten weeks overtime, since most trips were at weekends; on her return from each trip, she would immediately be plunged into planning for the next one.

More recently, the policy has changed again, with the recognition that experiences on the reserves help people understand and engage more passionately with conservation objectives.

But Blake is resilient and optimistic. Being thrown in at the deep end was, she believes, an amazing way to learn. In 2015, with ten years experience as access and visitation coordinator, she has brought order and system to a process that has been subject to different policy decisions. When she began, the policy was to encourage people out onto properties, providing gourmet lunches that she and other staff frantically put together beforehand. Later, the policy swung in the opposite direction, because visits overloaded reserve managers. More recently, the policy has changed again, with the recognition that experiences on the reserves help people understand and engage more passionately with conservation objectives. Blake has worked with all the changes, managing and monitoring every aspect of visits. She remembers the early years as 'insane, completely insane, but, in retrospect, the crazy chaotic times are the ones you remember most vividly'.[46]

Visits to properties have always been part of Bush Heritage's interaction with its major donors, supporters and volunteers. Nathan Males was in charge of the early trips in Tasmania, but with fifteen properties scattered over Australia, there was greater complexity in coordinating visits and designing programs to enhance visitors' experiences, without adding too much more to reserve managers' roles, particularly as many were new to the job. It was not quite the resource management position Blake had envisaged, but she found her rhythm and talent in developing a comprehensive visitation program, and arbitrating between the fundraising and marketing staff in Melbourne and the reserve managers, the two groups having different expectations as to what could and should be achieved. The addition of ten or more people on a property often strained the facilities in remote places, and put pressure on the reserve managers' time. Understanding the needs and limitations of both groups, and facilitating positive communication between them, Blake has been significant in building a stronger, more focused team.

Her first trip to Ethabuka Reserve with fourteen supporters remains memorable, mainly because 'everything that could go wrong did go wrong'. The news, delivered just before the trip, that reserve managers Karen Harrland and Al Dermer would not be there was the first intimation of difficulties. At 3 am on the first night on the property, Blake was woken from her sleep in an outside tin-roofed work shed (there was not enough room in the house for Bush Heritage employees) by a spectacular thunder and lightning show that lit the vast night sky continuously in a dazzling display while delivering torrential rain, 126 millimetres in forty-eight hours. The group was stranded, unable to tour the property or to leave. They were marooned at Ethabuka homestead for days, with limited water supply and no electricity, because the windmill that drove the generator had failed. Although there was food in the coolroom, it had to be rationed as it was impossible to know how long they would

Spectacular skies and a drowned desert meant memorable times for Bush Heritage staff and visitors, Ethabuka Reserve, Qld.

Photo: Julian Fennessy

Annette Stewart is congratulated by Luke Bayley (left) and David Whitelaw on receiving her Fulbright Scholarship.

Source: David Whitelaw

be cut off. The computers were down, there was no mobile phone access, and Blake had to rearrange schedules for guests who had other commitments by satellite phone, the only connection with the outside world.[47]

But it was also a rare opportunity to watch the desert blooming after ten years of drought, and to discover this in the company of ecologist Murray Haseler. He was a fount of information on the wildlife as they explored nearby sodden country bursting into life. Frogs emerged in their hundreds from years of waiting in the sandhills for this moment; fish came with the coursing streams filling waterholes; birds of every type flew in, attracted by hatching insects and fish, or trees and bushes glowing with flowers and new growth. The red sandhills were alive with small marsupials, lizards and other desert creatures that had hunkered down during the drought and were making the most of the unexpected rainfall.

For others based in the new office in Melbourne, there was the challenge of making sense of what was happening out in the field. Towards the end of this expansion period

The dramatic desert landcsape of Cravens Peak Reserve, Qld

Photo: Wayne Lawler/Ecopix

Annette Stewart, now office manager for a growing number of staff, finally turned her attention from trying to fix the business systems through the Finance Department to seeking solutions within the Conservation area. Initially uncomfortable with people who appeared set in their ways, she began to understand the difficulties they faced, realising that the program she was attempting to build would need to be different from any she had used before.

Although she had a Masters degree in Environmental Management and knew the theory, Stewart had little understanding of the practicalities of conservation. She had no contacts with other conservation groups, and no networks. So it was 'just luck' when, some months later, as she was searching the internet, she 'literally stumbled' across a system that, even in its imperfect state, she could see would suit her needs. There were:

> gold nuggets sitting there to be picked up. I could not believe it, I kept clicking on the next page and the next, and asking myself: Where is the catch? I contacted the people who were building it. They were up to version 2 and I got to know them. They could see I understood what they were trying to do. Since then I have been working closely with them.[48]

The group, comprising many of the world's leading conservation organisations and their key funders, known as the Conservation Measures Partnership, developed a methodology called 'Open Standards for the Practice of Conservation'. It would undergo many iterations, and gain in sophistication with input from users as they drilled down into the nitty gritty of the material. But Stewart took her time before customising the system for Bush Heritage.

The beauty of the system for its users is that, at any time, any question about funding allocations can be answered by a click.

In essence, the process was a systematic five-stage adaptive management cycle, covering every aspect of a project. Parts of the methodology were already being used around the world by other conservation groups, but, as Stewart observes, 'If you didn't have a systems background to put all the pieces together you could not see the beauty of it'. She brought it all together, and began to implement it at Bush Heritage, and to monitor its use. Open Standards was the means by which Bush Heritage became more efficient and transparent, by simplifying, better targeting and amalgamating management plans and evaluation processes for the best return for each dollar spent. This fed into more effective fundraising. But the process took time, and it would be 2012 before it was understood by everyone. Stewart demonstrated the system to other organisations, and it became an unexpected avenue for Bush Heritage's ambition to model an effective conservation methodology for others to follow.

The Miradi information management software that underpins the Open Standards system would eventually cover every aspect of Bush Heritage work. Because it was cautiously incorporated into each department, starting with those in the field, the whole picture was not immediately apparent. Introduced in this low-key way, it did not create too many expectations before providing results. Like most new systems, acceptance was gradual, and was achieved only after the benefits were understood. For Stewart, clarity was the objective:

> The conservation view of the world is we have to know what we are trying to do, what is it that we are trying to protect and what are the threats, and how can we intervene? The process and system we use has a language for these issues. The management plan for a reserve then becomes more directed towards dealing with the threats to the conservation asset. The plan and the budget are built around that, which is then hooked into the finance system, so that each property is regarded in a holistic way. This gives the fundraisers a good understanding of what we are trying to do, which they can then communicate to donors, so that right from the beginning, in our reporting, we can go to the Board to show what we are doing and why we are doing it.[49]

The beauty of the system for its users is that, at any time, any question about funding allocations can be answered with one click of the mouse. 'If you want to know how much of the budget for a particular property has been spent on eradicating weeds this year, it is there instantly.' The reward for Stewart came from a regional manager, who had been working in conservation for years. When he told Stewart that he finally had a system that met his needs, 'that was a real tick for me because the system is creating a more efficient workforce'.[50]

Stewart and her group spent several years refining the system, and the improvements she developed have now been adopted and incorporated into the software at Bush Heritage, and are being used by conservation organisations throughout the world. Her work makes her one of Bush Heritage's heroines, but her shy demeanour gives none of this away. Only when prompted, will she admit her pride in the system. Others say that Stewart's work underpins the whole organisation as it develops new strategies and copes with almost exponential growth. But Stewart, with a vision for what could be achieved, wants more from the system, despite recent admiring comments from the chief information officer of one of the US national park services, who knows that the same kind of day-to-day transparency of activity and financial spending is difficult to achieve in government organisations. In 2015 Stewart won the only Fulbright Scholarship awarded in the Australian charity sector. She spent six months in the US to work with others to develop the system to its full potential.

At the end of 2006, Board members and senior management decided it was time to revitalise Bush Heritage's name and logo to reflect the organisation's new dynamism and fresh approach to the protection of Australian landscapes, and in March 2007 the changes were made public. Since Bush Heritage could now truly call itself a national organisation, it was decided to drop the word 'Fund' from its title (Australian Bush Heritage Fund), which had always been confusing, in favour of Bush Heritage Australia, since most people already referred to it simply as Bush Heritage. Alexandra de Blas, Bush Heritage media strategist, who had been a well-known voice on ABC Radio National's *Earthbeat* for eight years before she came to Bush Heritage, managed the rebranding of the organisation, and helped build Bush Heritage's public profile and media presence in the period between 2005 and 2008. As part of this strategy, she overhauled and re-created the Bush Heritage website, and dramatically increased the organisation's media exposure. She also considered the logo that had been designed in 1990 to symbolise Bush Heritage's activities. Changing it was a significant step forward in expressing the expanded vision. The original three tall eucalypts were transformed into three horizontal brushstrokes in brown, ochre and blue, colours symbolising a classic Australian landscape – the rich, warm colours of the earth, and the blue of the water and the vast outback skies. As Toyne commented in the autumn newsletter: '[This] more appropriately represents our new way of thinking about conservation; we are no longer working just to protect single properties but to conserve whole landscapes of land and water'.[51]

The new logo colours reflect the Australian landscape.

BUSH HERITAGE
AUSTRALIA

The one that got away

In 2007 the Board was also considering its most momentous purchase, Wollogorang–Wentworth, two pastoral leases on the Queensland/Northern Territory border. The properties totalled 705 000 hectares, encompassing Gulf Coastal, Gulf Plains and Gulf Fall and Uplands bioregions in northern Australia.

TOGETHER THE PROPERTIES would be an exceptional acquisition that met all Bush Heritage's goals for protecting threatened ecosystems, species and significant and highly productive landscapes, all within a Bush Heritage anchor region. At the time, grazing cattle were degrading both properties and inappropriate fire regimes were exacerbating the spread of weeds and erosion, threatening the future of several key species.

The properties had several potential income sources, including light grazing, tourism, training and education in environmental management, particularly for Aboriginal groups, and carbon and biodiversity credits. That these opportunities were considered indicates the seismic shift in Bush Heritage thinking, highlighting the underlying themes embedded in both the Beyond the Boundaries and Anchors in the Landscape programs. Local Aboriginal communities still had strong links with their country and Bush Heritage saw opportunities to develop meaningful co-management approaches, with the possible long-term goal of handing back title to the Traditional Owners, who had indicated a desire to take on ownership if and when circumstances were right.

The purchase had exercised the collective mind of the Board for some time. Cowell, Toyne and Humann had all explored the country, and Humann and Toyne both spoke about it later with an intensity of longing. Humann recalled the glory, the range of magnificent landscapes, its remoteness: 'The property was huge, and ran from high escarpments into the Gulf, beautiful bushland, spectacular, with interesting fauna and flora and extraordinary landscapes'.[52] The asking price was between $30 million and $50 million, depending how the deal was struck. Humann and Fitzhardinge initiated discussion with the owner, and the Board authorised the development of a full business case. A long-term loan from TNC was possible, and Toyne and Humann met with Peter Garrett, the federal Minister for the Environment, who was receptive to the idea. They also had a multi-million dollar pledge from an overseas donor who had visited the property with them.

There were management issues. No one wanted to buy the 30 000 head of mixed breed scrub cattle. Cowell considered the possibility of Bush Heritage running the cattle, and working with the Indigenous Land Corporation to develop an Aboriginal enterprise, but this was difficult to sort out in time to convince the Board and supporters. Toyne thought it would be a wonderful achievement, and wanted to take the risk.

Humann was in Vancouver, finalising arrangements with TNC for a long-term loan for the property, when he heard of the Lehmann Brothers collapse, the first sign of global financial instability. In that climate, dependent as Bush Heritage was on private and government funding, purchasing the property could have been a disaster. And the Board decided against it.[53] That decision turned out to be prudent, because the Western world was on the cusp of the Global Financial Crisis (GFC) that began in 2007, and came to a head in October 2008. Staff sighed with relief, and Humann had to tell the overseas donor that the deal had collapsed.

The donor was disappointed, but Humann pointed out what could still be achieved. Having been particularly interested in the connections that Bush Heritage would have made with the local Aboriginal communities on the Wollogorang–Wentworth properties, the donor soon realised that his donation could instead support existing Aboriginal partnerships that Bush Heritage was developing in the far north. And, despite his regrets, Humann found satisfaction in helping with the transition from the buzz of land purchase to funding less obvious elements of the program.

The failure to buy Wollogorang signalled the end of Bush Heritage's expansionist period. Although buying property was one of the most exciting aspects of the program, and one that donors almost invariably supported, many at Bush Heritage believed it was time to consolidate, and to plan for the long term, since the goal was to conserve and protect land forever. Perpetuity is a seriously big concept, and it would require a little more finessing than had been previously demonstrated, if Bush Heritage was to maintain this goal.

Wollogorang-Wentworth was considered one of the most important properties to protect.

Photo: Kate Fitzherbert

PART 4

While the GFC put the brakes on land purchases, it allowed Bush Heritage to focus on managing existing reserves, and on establishing programs to conserve land in perpetuity. Bush Heritage embarked on several ambitious projects, raising donor awareness and excitement, and giving it a new impetus to continue as enthusiastically and passionately as it had begun.

Preparing for perpetuity

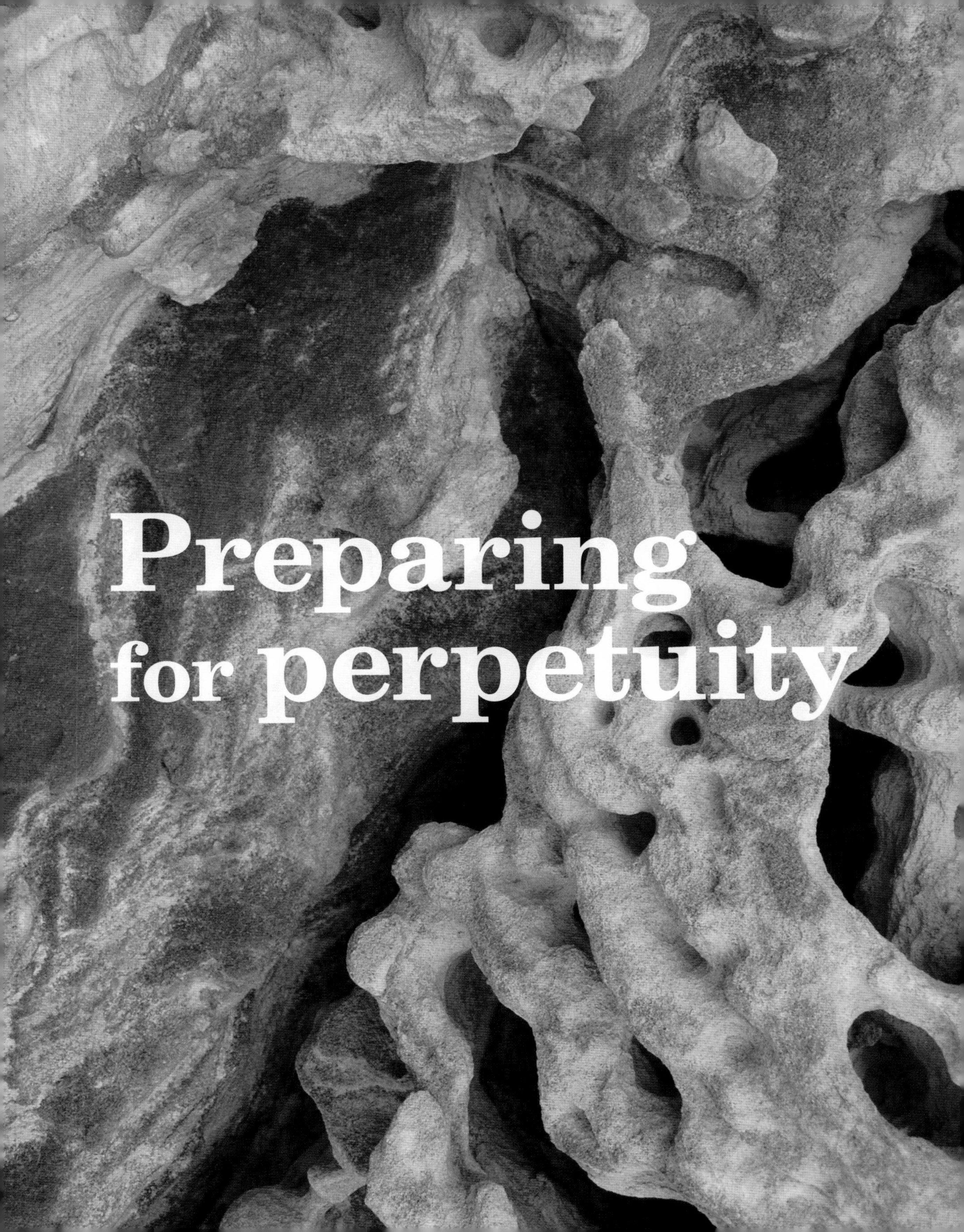

A crisis of confidence

At the end of 2008, when Wollogorang was lost, the first stage of the Anchors in the Landscape program officially ended. The marketing campaign had been successful, achieving a total of $23 million, with twelve properties acquired. However, the failure to buy Wollogorang had repercussions.

THOSE WHO HAD WORKED to make it a reality were devastated. Most of the responsibility for putting together realistic figures for the project had fallen on Stuart Cowell. The work, and its failure to convince, took its toll. He left the organisation in 2010, burnt out by his workload and dispirited by what still had to be done.[1] Others were relieved. Louise Sylvan, who succeeded Toyne as President in November 2009, felt the purchase would have bankrupted Bush Heritage.[2] Whatever the feelings, the end of the affair, and the Global Financial Crisis (GFC), heralded a numbness that gripped the administration for several years. The Board adopted a prudent 'wait and see' approach, anticipating that the global financial situation might affect donation income, and unwilling to increase expenditure until it was sure of the impacts. In this climate, the organisation was restructured, resulting in some staff reductions. Don Bell, who had managed Aboriginal partnerships on the reserves, was a casualty, since the view now was that reserve managers themselves should pursue Aboriginal partnership possibilities on their reserves.

However, despite fears that the GFC would adversely affect donation income, Bush Heritage's major gifts income rose in the 2008–09 year, with an overall increase in donations. But this could not compensate for the decline in government funding from the National Reserve System (NRS) for land purchases. With an expanded suite of properties, Bush Heritage had to do more with less money, uncertain if and when the financial pain would end. In this atmosphere, staff came under increasing pressure, particularly as planning was already underway for the next phase of the Anchors in the Landscape program scheduled to begin in 2009.

But not all the repercussions from the Wollogorang affair were bad. One result was that Board members began to focus more on long-term strategies, which could only be achieved with a solid administrative structure to support them. At the time this structure was not in place.

Once again, too few staff were trying to fulfil too many roles. Humann was overseeing the purchase of properties, and their management, developing fundraising strategies, and signing off on the appointment of new staff, including those living in remote places. This latter aspect on its own was almost a whole new field of endeavour for Bush Heritage. Given that workload, the organisational structure, with Humann at the top, was stretched beyond capacity, sagging under its own weight. Volunteer caretakers looked after properties until reserve managers could be appointed, stop-gap measures that often lasted a year. While the volunteers made valuable personal contributions, often through building and repairing infrastructure, it was not a strategy that enhanced Bush Heritage's credibility with donors and other supporters.

The GFC forced directional change on Bush Heritage. The puff went out of buying and marketing the next acquisition to supporters, and newsletters began to focus on work on the reserves. Not enough, according to some, who felt more care was needed to assess the state of a property immediately after purchase. For example, a BioBlitz, in which ecologists, botanists, zoologists, entomologists, and other experts and volunteers could spend a fortnight capturing and recording finds, would help create a baseline for further scientific research into existing species' prevalence and health. This process would be adopted with later purchases, but in 2009 it was simply too expensive. As a pastoral landholder, Guy Fitzhardinge was critical of Bush Heritage's purchasing agenda, believing that more resources should be devoted to management of the existing reserves. On resigning from the Board, he suggested appointing Mike Chuk as his replacement. Chuk was working in natural resource management for Desert Channels Queensland and had practical experience in reserve management planning, fire management and invasive weed control.

Plans went ahead to amalgamate Communications and Fundraising into an integrated Marketing unit of three teams: Fundraising and Communications; Supporter Services, to provide supporters with more efficient responses and receipting; and Community Engagement. In this latter area the newly developed Bush Heritage website became a major tool in disseminating information for volunteers and supporters keen to offer help and expertise on administrative tasks or on reserves.

At the heart of this restructuring was Toyne's desire to spread the Bush Heritage message further afield to attract more donors and volunteers. Additional staff with different skills were required to build fundraising and marketing capacity, but to some of the reserve managers, struggling with ambitious management plans and few resources, these efforts seemed to be directed at the glossy side of the business. Some left, disillusioned, unsure how necessary these aspects were to the core work of conservation.

But the Board was strengthening the Conservation arm, too. A chief conservation manager was appointed in 2008, with responsibility for developing strategies for reserve management, overseeing reserve managers, developing science policies and managing Aboriginal partnerships. For this role, Humann sought someone with a science background and people management skills to complement his focus on the stewardship of supporters, engagement with new donors, oversight of the financial management of the organisation, and assessment of properties. He appointed Nicola Markus.

Markus gained a PhD in Zoology through studying flying foxes. She had been the Australian manager of Species Programs with World Wide Fund for Nature (WWF), and dealing with scientific reports from scientists of all persuasions had inspired her to write about current political thinking and its effect on environmental policy. *On Our Watch: The race to save Australia's environment* is a synthesis of scientific and political views and activities relating to the environment in Australia, and a passionate plea for support and involvement in environmental outcomes.[3] She was also keen to develop partnerships with Aboriginal people, an activity that was emerging as significant in the conservation program, and garnering donor support. When she joined Bush Heritage, Markus believed that the acquisition of the Wollogorang–Wentworth property would have stretched Bush Heritage beyond its capacity.

Buying property is tangible. Less clear-cut is the ongoing process of managing that property. Some still believed that, after the expensive key recovery steps in the first few years, management of a property would fade into ongoing maintenance, reducing the financial drain and employment needs. But this was a fallacy. Ongoing scientific research would uncover new problems that had to be addressed; more accurate quantifying of properties' conservation values would lead to further work. Ecologists, remembering cane toads and the gambusia approach to mosquito control, wanted to ensure that one ecological solution did not create other problems, so that more research and trial sites with appropriate evaluation methodology were always necessary. Budgets would therefore have to be much greater than had been foreseen, and would continue to be so.

The restructuring saw Ray O'Halloran, who had earlier been involved in developing the safety system, become operations manager, while Markus, as chief conservation manager, worked alongside Humann, giving him more time to deal with donors, staff appointments, government regulations and property assessments. When O'Halloran left, Dave Whitelaw, who had worked in several government environmental departments, took the position. With Whitelaw's arrival, Markus had overarching responsibility for three teams: Kate Fitzherbert, the assessment and property services manager, looked after a broad portfolio that took in the growing number of ecologists; Stuart Cowell worked with the Aboriginal and property partnerships; and Whitelaw was in charge of all the reserves and reserve management. The four met regularly before Markus reported back to the group comprising Humann, the finance manager and the chief marketing manager.

Health and safety is one way of gauging a company's commitment to its employees and to a long-term future. With a senior management group now in place, and an interim level of regional leaders planned, it was time to embed the Health and Safety Management System into the culture of the organisation, and senior management took on this responsibility. Al Dermer, Ethabuka's reserve manager, brought the concept of the 'grab bag' to Bush Heritage from his experience in the Australian Antarctic Division, and Whitelaw was charged with developing an appropriate model for Bush Heritage. It soon became mandatory for every person, or group, including volunteers, to carry a grab bag, containing a medical kit, a satellite phone and an emergency Global Positioning System (GPS) tracking device (Spot Locator) that could be used to indicate position, and to emit an emergency call via satellite to the central call centre to be immediately acted upon. Whitelaw displayed the contents of the bag at a Board meeting in 2010, impressing on his audience the need to make the bag part of the everyday life of people working in the field, so that working safely gradually engrained itself into Bush Heritage culture. The grab bag was included alongside the basic rules of always carrying a shovel and an extra jerrycan of water in the vehicle, and remaining with the vehicle in the event of a breakdown. It was also now mandatory to log in to an independent safety call centre and to log out at the end of the working day. The clearly defined chain of responsibility from fieldworker to regional manager to senior management to CEO would be activated at any divergence from accepted practice, underlining just how seriously Bush Heritage took the safety of its employees. It took time, but by 2011 these procedures were accepted as normal practice and a useful safeguard. This was later reinforced by safety specialist company, and pro bono supporter, Sentis. Since the company's association with Bush Heritage began in 2013, it has presented at Bush Heritage's biennial Gatherings, working with staff through enjoyable and mind-opening sessions to promote and develop the 'culture of safety'.

Creating the senior management team was an early attempt at decentralisation, and recognition that different skills were needed to achieve success and longevity in conservation work. Although there was always a noticeable regeneration of the natural vegetation after destocking a new property, Bush Heritage's claim to integrity and management of threatened species over time would always require serious weed work, serious fire work and ongoing feral management. As Markus remarked: 'It took 100 to 200 years to ruin the land, and it was not going to be fixed in five years'.[4]

The division of roles suited Markus and Whitelaw's complementary areas of expertise; the latter's hands-on practical approach was particularly appreciated by the reserve managers. On appointing Whitelaw, Humann advised him that Bush Heritage was not 'government' and that he did not expect him to act as a public servant, commenting that the team in Melbourne was doing fine work and it was important that the team on the ground, the reserve managers, realised this.

By the end of 2008, Bush Heritage had purchased thirty properties dotted around Australia, many in remote locations and requiring permanent on-site managers.

For Whitelaw his role as operations manager was a dream job. He visited all the properties that Bush Heritage was managing, meeting 'amazingly fantastic, highly confident, passionate people', who needed to be reassured that they were doing a great job, and that the Melbourne office would resolve the problems they were experiencing.[5] Clearly, a cultural divide had been developing between the national office and those in the field. Not so long before, Steve Heggie and Mel Sheppard, Carnarvon's first reserve managers, had believed that the organisation existed entirely for them, catering to every desire they had for the property. By 2008 the lines of communication had become so convoluted that it took several approvals up the hierarchy for a reserve manager to buy a coil of wire. What had changed?

By the end of 2008, Bush Heritage had purchased thirty properties dotted around Australia, many in remote locations and requiring permanent on-site managers. Each property had individual needs and management issues that could not be treated generically. Communication between staff in the central office and those in remote corners of Australia was difficult, and in 2008 Annette Stewart's work on conservation systems had only just begun. Money was tight, as always, and while the Melbourne office had grown substantially to keep up with the demands being made on the organisation, it was fragmented into departments, and links between those in the office and those in the field were more tenuous. Moreover, as Markus had predicted, the Board was coming to terms with the fact that successful management of each property required far more money than had at first been budgeted. This meant that an integrated program of publicity and fundraising to support each reserve manager's work and highlight each property's individual character was necessary: most of this work would derive from the national office in Melbourne, and it was imperative that the two arms of the organisation synchronise their activities.

The people on the ground

Employing reserve managers was necessarily occupying more of Humann's time. Not every applicant who looked good on paper and who interviewed well was actually suited to the job. There have been many reserve managers on Bush Heritage properties in its twenty-five years, too many to mention here.

BY FOCUSING ON SEVERAL who have found an ongoing career in Bush Heritage, it is possible to gain an insight into how Bush Heritage found its way towards a stronger management position that best provides for its distant employees and supports its long-term vision.

Reserve managers are usually, but not always, people with practical bush experience, coming from positions with state or federal governments in national parks, or from farming backgrounds. They find working for an NGO 'liberating'. They enjoy being their own boss, responsible for the property they are managing and finding innovative solutions to problems without an intervening bureaucracy, and they can cope in extreme weather conditions. They rejoice in the sights and sounds and smells of the bush, the seasonal changes, the glimpses of shy animals, of birds and insects, the discovery of new growth on old trees, and new plants, the rich, full silence of being alone and at one with the landscape. Most of all, they are committed to conservation, and believe that what they are doing is important. They are patient, and recognise that their work is part of a long-term strategy, and that they will not see the full maturing of the projects that engage them. They would probably have agreed with theologian Davis McCaughey, who would become Governor of Victoria in 1986, when he addressed an agricultural college ceremony in Western Australia in 1984: 'If I love the land, or love learning, I want to leave it better able to serve the future generation than I found it. I do not wish to violate its nature. There is an art of living and a great part of its practice depends on respect; and respect itself is a part of true love. In conservation the art and science of farming come together to serve a noble cause'.[6] This was an early public expression of the connection between human wellbeing and the natural environment. Two years later the National Forum on BioDiversity was held in Washington, DC, and in the collection of significant scientific papers presented by conservation scientists and edited by EO Wilson, the psychological need of humans for interaction with the natural world was also recognised.[7]

Humans are grounded and brought back to reality through the 'wonder and excitement and challenge' of the natural elements, and empathise with the idea that 'long-established existence somehow confers a right to a continued existence', a sentiment that can be experienced in the silence that envelops gnarled trees and rock cliffs mirrored in still water.[8] Andrew Harper, founder and leader of the travel firm Australian Desert Expeditions, believes that 'when you sleep in the open in the desert under a full moon it attaches a thread to your heart, and keeps pulling you back'.[9] A reserve manager, although perhaps not prepared to express the spiritual and aesthetic pleasures of being at home in the Australian bush, is likely to sense that inexplicable pull of the land that brings inner joy. And, having lived remotely and revelled in that environment, the return to urban life might never be quite enough.

Steve Heggie and Mel Sheppard knew they were pioneering a new way of life, and they were lucky in ways that no longer exist for Carnarvon Station Reserve. They had two mail deliveries a week, and 'fresh cos lettuce and other delicacies' could be delivered with the mail by phoning in.[10] They built large fuel tanks to cater for fuel deliveries once a month. These services disappeared as the communities in the area dwindled, and managers today have to drive the round trip of 400 kilometres to fill their mobile fuel tanks.

Heggie and Sheppard were responsible for rounding up the last rogue cattle on Carnarvon. Mick Blackman, the next manager, would do work on a fire management plan, eagerly awaited at head office, to be signed off by the conservation managers and the Board for implementation. As mentioned earlier, he also installed a secure water system for the homestead dwellings. When Darren Larcombe arrived in 2004 as Carnarvon's third manager, after working as a ranger at Watarrka National Park (in which King's Canyon is located) in the Northern Territory, the hard yards had been done, and neighbours were more relaxed about the idea of 'a whole bunch of greenies living up the river'.[11] The management plan was in place so that projects could begin. By this time, volunteers, mostly Bush Heritage supporters, were flocking to Carnarvon, keen to work in the bush.

A reserve manager is likely to sense that inexplicable pull of the land that brings inner joy.

Heggie and Sheppard, and Blackman and his wife Clare, all initiated contact with the Bidjara people; Clare particularly worked hard on the relationship. At first it was difficult, because until 2001 the Bidjara people had been denied access to Carnarvon, and only now were they beginning to reconnect with this part of their traditional lands. They visited the caves with rock art, etchings and engravings, and the ochre pits of deep purple and crimson ochre, and began slowly to share their stories of the land with the reserve managers. By Larcombe's time the Indigenous Land Corporation had already acquired Mt Tabor Station on behalf of the Bidjara people. The Traditional Owners were planning to protect the rock art, mitigate feral animal impact and undertake weed control, and they worked with supporters on Carnarvon to build fences to stop wild horses damaging the important sites. Larcombe oversaw the beginning of a more direct relationship with the Traditional Owners when a Bidjara man became a staff member, employed on a scholarship established to honour Rick Farley. He joined the team at Carnarvon as a trainee in conservation management. The scholarship was a positive initiative, but its success at Carnarvon was compromised by a lack of support structures for both the appointee and the reserve manager.

Larcombe recommended that volunteers stay at least two weeks, giving time to induct them into safety programs and into managing tools so that they were able to participate in good conservation outcomes. In his first year as manager, there were only three nights when Carnarvon did not have someone – volunteers, visitors, or Bush Heritage staff – on the property. Despite its remoteness, Carnarvon was buzzing. A generous donation bought a seed harvester, and Larcombe was able to seed native grasses on the denuded plains. Hills and gully country had been damaged by wild horses, and earth-moving facilities helped 'adjust the landscape' for grass seeding, to halt serious erosion.

Glen Norris found his world changed by a seven-week stint as a volunteer at Carnarvon with Larcombe, who was a 'fantastic manager who really knew his stuff and taught me a lot'.[12] Norris had left school feeling he had learnt nothing, and fallen into a job as an earth-mover, driving bulldozers and bobcats. Although the money was good, at the age of twenty-seven he decided to return to study, doing a Bachelor of Social Sciences course at RMIT in Melbourne that took in sociology, social sciences, statistics, environmental resource management and earth management. It gave him a new love of learning. At the end of his first year he went to Carnarvon and 'fell in love with the place, absolutely'. He applied for a job as Bush Heritage operations officer, and completed his degree part-time, alongside full-time work.

Working in the Bush Heritage Melbourne office, Norris was on a steep learning curve, and keen to learn as much as he could. But after three years he yearned to get back to the bush, and in 2007 applied, unsuccessfully, for the reserve manager job at Scottsdale, near Canberra. In retrospect he realises that he was not right for the job and that, for him, it was not quite 'outback' enough.

But Bon Bon Station, in South Australia, acquired in 2008, also needed a manager. It was 213 000 hectares of native vegetation in semi-arid country, outback in the raw. Norris drove to Bon Bon with his dog, surveyed the landscape, taking in the dryness, the wide, open spaces and the silences, and knew, like Bob Brown in Tasmania, that he had 'come home'. He threw himself into the work, his practical skills, the years of knocking about the country as an earth-mover, and his capacity for friendship fitting him for the job.

Friendship with the conservative cattlemen and sheep farmers in arid regions was not a given, however. They viewed with distrust the 'greenie' who had no stock and no family. When he arrived at a Christmas party held at a neighbouring station with a couple of zesty lemon tarts and a batch of blueberry muffins he had cooked himself, despite being told to 'just bring a swag and an esky', the

Glen Norris and Bush Heritage media manager Annette Ruzicka

Source: Annette Ruzicka

tough men of the region jumped to the erroneous conclusion that he was gay. But the women were impressed with his culinary contributions, and he gradually broke down the barriers, convincing even the staunchest pastoralist that he was one of them, that they all had the same aspirations, whether they were growing sheep, cattle, grain or native vegetation. And at gatherings in the future, everyone looked forward to Norris's blueberry muffins.

Norris also made connections with the Antakirinja Matu-Yankunytjatjara people who maintain strong links to their land, since Aboriginal families had worked on the property until the mid-1970s. Like Phillip Toyne who had earlier worked as a lawyer in this area, he learnt of the *tjukurpa*, the creation period when ancestral beings (*Tjukaritja*) created the world as it is today, and the law for relationships between people, plants, animals and the land. The Traditional Owners had been fighting to have their rights, interests and responsibilities for their country acknowledged, but it would not be until 2011 that their rights to hunt, fish, live, camp, and gather and use the natural resources, undertake cultural activities, and protect places of cultural significance were officially recognised. Norris began the relationship that resulted in a collaborative management arrangement with Traditional Owners, bringing Aboriginal people back to Bon Bon in 2014, building trust in the 'two-way' discussion of shared concern for the land. This led to a cultural heritage survey, in collaboration with the SA Museum, pinpointing areas of Aboriginal cultural sensitivity to be integrated into the reserve management plan, to prevent damage to the cultural sites.

During his time at Bon Bon, the rains came. Lake Puckridge, the centre of a huge wetland, filled with water, attracting migratory birds and boosting growth in the endemic species of mallee on the dunes and margins of the lakes. Norris explored the country, awash with brilliant flowering plants, a hundred different kinds of *Eremophila* among others, abounding bird and animal life, and was happy. But this ended suddenly. One day, travelling the Stuart Highway, he stopped to help a man whose trayback truck had broken down. The tray was up, and Norris was investigating the problem when the tray came down, pinning him by the arm. Badly injured, he was airlifted to hospital, and the months of recovery made it difficult to return to his remote paradise.

Mike Chuk, whom Guy Fitzhardinge had encouraged onto the Board in 2010, had been considering a change in his working life. He and his wife Julia had earlier applied to become reserve managers at Bon Bon, and were living in Longreach in Queensland when Norris was taken to hospital. Chuk had visited Bon Bon in the past and knew he would like living there, and he and Julia were looking for something they could do together. Norris's accident meant that Bon Bon was suddenly without a manager. When contacted by Bush Heritage, they made a quick decision, packed their belongings and headed off.

Norris began the relationship that resulted in a collaborative management arrangement with Traditional Owners, bringing Aboriginal people back to Bon Bon in 2014.

Chuk's experiences fitted him both physically and emotionally for a reserve manager position. A childhood in the bush on the outskirts of Ballarat, and an interest in birds were followed by a degree in Forestry and Invertebrate Zoology at the Australian National University in Canberra. After completing his studies, he travelled north, working on locust control campaigns in southern Queensland, in the Channel Country and in the far north of South Australia, counting himself lucky to have explored these regions before tourism encroached. He enjoyed years in Broken Hill before heading up to the Northern Territory where he was the first ranger at Watarrka National Park (Darren Larcombe would follow later). He was eleven years in the Northern Territory before taking up a position with Desert Channels Queensland, which involved him in consultancy work with Bush Heritage. Increasingly, his interests were turning to natural resource management, and he helped set up the first reserve management plan for Cravens Peak and a fire management plan for Edgbaston, at which point he was recruited onto the Bush Heritage Board.

Julia Harris, Chuk's wife, also had a love of the bush developed in childhood. Her educational path took her into Agricultural Science. A job in the sheep farming industry alerted her to the fact that sustainable production in the pastoral zone depended on controlling weeds and feral animals. She met Chuk at Longreach, their shared interest in conservation and volunteering bringing them together, and eventually to Bush Heritage.

Bush Heritage reserve managers and ecologists are on the frontline in the battle to save the Australian bush, and they can change mindsets by showing people the properties.

Chuk resigned from the Board to take up the position as reserve manager at Bon Bon. He and Harris took on the job as a couple, as Heggie and Sheppard had done, so that they could work together, pooling their complementary skills. His deep and encyclopaedic understanding of conservation issues and strategy underpins her more practical experience of fencing, driving tractors, and managing weeds and feral animals. Harris's willingness to tackle the reporting and budgetary demands that managing a conservation property imposes leaves Chuk free to explore the scientific and technical aspects of conservation that benefit Bon Bon. Together they revel in the wild beauty of their domain – the dry salt lakes shimmering silver and ringed by pink samphire bushes tinged deep ruby in the setting sun; in the Hidden Valley, the weird massive structures built by resident rare hairy-nosed wombats; night-time tracks of small marsupials, lizards and snakes in the red sand creekbeds, discovered in the morning after a night of camping under the stars.

Because the Stuart Highway cuts through Bon Bon, making it readily accessible, the property attracts a steady stream of visitors. Chuk and Harris found, as have other reserve managers, that their conservation and practical farming skills are only part of the job, and they entered wholeheartedly into the life of the far-flung community in which they live. One big step towards a higher profile in the area was their participation in the camp oven cook-off, a competition dreamt up by the enterprising publican at Kingoonyah pub to encourage tourism. Each year he welcomes the entrants and chooses the judging panel from customers on the night. The event has become a rallying point, with people travelling hundreds of kilometres to join in. Competitors bring their own camp oven, are supplied with meat, and can choose other ingredients from the pantry to create their dish. When the Bon Bon team entered the competition, Chuk asked whether contributions from the bar were included, and, on being given an affirmative, promptly poured a bottle of port into the Bon Bon offering. Not surprisingly, their dish was judged the most delicious by the three judges, for a time turning the new managers at Bon Bon into local celebrities.

A less glamorous task facing Chuck and Harris is their ongoing battle with buffel grass, thought to have originally arrived in an Afghan camel harness 150 years ago. By 1910 the seed was being actively distributed by departments of Agriculture as a pasture grass. Quickly becoming the pasture grass of choice in the north, it has spread across the country, destroying bushland because it fuels devastatingly hot wildfires. It is threatening the health of Bon Bon and, since 2013, Chuck has been systematically spraying the invasive grass to bring the infestations under control, in a program that has involved Bush Heritage volunteers, SA Main Roads, the SA Arid Lands Natural Resource Management Board and the Native Vegetation Council. Overall, 65 kilometres along the Stuart Highway have been sprayed, and a further 55 kilometres of minor roads and creek lines. It is only by systematic and consistent monitoring and treatment over many years that the scourge will be eradicated. But the problem is one that affects many regions, and Bush Heritage has embarked on a new science- and research-

Paul and Leanne Hales, first reserve managers at Eurardy, WA, who moved across the continent with their family to manage Yourka Reserve, Qld

Source: Paul Foreman

based program to better understand the impact of buffel grass on biodiversity, enlisting students from South Australia's Flinders University to join arid country ecologist Aaron Fenner in the research.

Bush Heritage reserve managers and ecologists are on the frontline in the battle to save the Australian bush, and they can change mindsets by showing people the properties and describing the failures and successes of practical conservation. Leanne and Paul Hales discovered this when they became Eurardy's first reserve managers in 2005. Because Eurardy had for many years been a favourite camping stopover between Perth and Shark Bay in Western Australia, they had many opportunities to 'talk conservation' with visitors.

Both had worked with the Queensland Parks and Wildlife Service. Paul, from a farm background, was a ranger, and Leanne, a 'city kid', found her passion as a youth worker, running programs in the bush for teenagers at risk. While they enjoyed their jobs, they were often away in the field, but not together. The job in Western Australia would change that, and they set off in their car to a new life.

While crossing the Nullabor, Leanne realised she was pregnant. It seemed a disaster, and they debated whether to turn back. Karen Harrland and her partner Al Dermer had had the same reaction a year earlier when they learnt of Harrland's pregnancy just after they had moved to Ethabuka as its first reserve managers. Harrland's award-winning novel, *Spinifex Baby*, draws on her experience of having her first baby in the comparative isolation of the desert environment.[13] The Hales, too, realised that they could manage, and continued their journey.

There were hard times – refurbishing the infrastructure, building a reliable water supply, fixing fences, maintaining an open-gate policy which meant being inundated with visitors. They worked on a management plan, applied for funding for a wildflower tourism project to develop with the local Aboriginal community, and planned a self-guided interpretive circuit that took in the adjoining Kalbarri National Park. While Bush Heritage was deciding how to deal with the property, they maintained the agistment arrangements with a local sheep farmer, and leased several paddocks for wheat cropping, while leaving the bulk of the property for conservation. They communicated their love for Eurardy to visitors who dropped in and stayed, and returned again, many becoming Bush Heritage supporters and volunteers. One couple, impressed with the property's combination of commercial operations alongside conservation efforts, commented on departure that they (the Hales) had secured a million dollar bequest for Bush Heritage because of their practical approach. The visitors, farmers themselves, told them that, not being keen on 'greenies tying up the bush', they had arrived critical, but were leaving convinced that Bush Heritage was on the right track.

But an awkward situation loomed. Stuart Cowell and Sarah Eccles had already encountered anger from Aboriginal people who believed that Bush Heritage ought to hand land back. This became an issue for Eurardy while the Hales were reserve managers. Both Cowell and Humann had tried to identify the Traditional Owners before the purchase but were unsuccessful. After the sale was completed representatives from the nearby Aboriginal community protested that they had not been given the opportunity to buy the property. This caused considerable friction between the two parties. The incident reinforced the determination of Bush Heritage to consult prior to any purchase and not acquire land in competition with Aboriginal interests.

Perhaps this experience helped the Hales decide to return to northern Queensland, closer to their families and to familiar tropical weather, although they had grown to love the dry country. But they also loved the Bush Heritage culture, and wanted to remain as part of the Bush Heritage 'family'. Resolving their indecision was Bush Heritage's advertisement in 2007 for a reserve manager at the newly acquired Yourka, 130 kilometres south of Cairns. They could live in Kuranda, a town on the tableland, two hours from Yourka. So it was with excitement that they packed their bags again and journeyed back across the country to another Bush Heritage project.

Work on the reserves

After the purchase of a property, the hard work begins: developing management plans, implementing them and reporting on progress. Reserve managers and ecologists needed to establish baselines to understand the conservation values present at the time of handover, so that they can monitor their programs for subsequent outcomes.

ECOLOGISTS PROVIDED THE scientific guidance and planning for these activities, and the relationship between the reserve manager and the assigned ecologist was an important one. Trust and camaraderie, and a shared enjoyment of long days and nights in remote places, was built up over time and through repeat visits by the ecologist, who generally was responsible for several properties.

Ecologists need a wide range of scientific knowledge to deal with complex issues. Sandy Gilmore, an ecologist appointed in 2005, has worked on most of Bush Heritage's properties. His specialist interest is bird communities, and his practical interest is integrating ecology with land management. He is always excited when a rare or threatened species of animal, bird or plant is returning to its previous haunts, but also believes that existing common species should not be overlooked, because their function in maintaining a healthy ecology is essential. An illustration of this occurred for him at Bon Bon Station Reserve in 2014, when a large influx of masked woodswallows, not normally present on Bon Bon, settled temporarily to voraciously feed on grasshoppers that covered the ground, also a rare occurrence. The food bonanza caused the birds to move in on the resident black-faced woodswallows' habitat and nest in the dead mulga trees. Gilmore's calculation put the bird population at 500 000. His point is that, just because the grasshoppers and both species of bird are common, 'doesn't mean we shouldn't be concerned about their numbers – to maintain a functional ecosystem, it's very important that they remain common'.[14]

Fundraising staff in the Melbourne office knew that the scientific rigor and methodology that ecologists and reserve managers were applying was important to their side of the operation, too. Donations from The Nature Conservancy (TNC) and the Vincent Fairfax Family Foundation's funding for Scottsdale necessitated five-year ecological acquittal reports, and effective stewardship of individual donors also required the ability to convey to supporters the reality of scientific and conservation activity on reserves.

Between 2007 and 2010, substantial funding from the Macquarie Group supported an Ecological Outcomes and Monitoring Program, which financed the basic methodology established at the time. Eventually the monitoring program was integrated into the Miradi Planning System, but before this, conservation manager Paul Foreman, senior ecologist Jim Radford, and fundraiser Roewen Wishart collaborated on a program to measure changes on the properties over five years that would fulfil the requirements of donors such as TNC, and the Vincent Fairfax Family Foundation.

A collaboration between Bush Heritage and the Royal Geographical Society of Queensland brought together a multi-disciplinary team of thirty-two scientists and independent researchers for a comprehensive scientific study.

Bush Heritage ecologists have a broad understanding of all ecosystems, but no one is an expert on every aspect. Connections with local university groups and PhD students working on Bush Heritage properties is therefore important, and Bush Heritage had long been encouraging scientific research and developing collaborative partnerships wherever possible. Field research on Bush Heritage properties was often part of PhD theses or postdoctoral papers that complemented reserve managers' and ecologists' activities. For instance, a collaboration between Bush Heritage and the Royal Geographical Society of Queensland brought together a multi-disciplinary team of thirty-two scientists and independent researchers for a comprehensive scientific study soon after Bush Heritage bought Cravens Peak, and after rain had encouraged plants and animals to flourish. Projects ranged from studying bats and jumping spiders to tracing aquatic ecosystems and unearthing fossil fish. The data was recorded in on-site inventories of fauna and flora and contributed to ongoing monitoring, while helping to foreshadow further site surveys and fieldwork. And in Western Australia, lecturers, postgraduates and students from Edith Cowan University's Centre for Ecosystem Management regularly visit Charles Darwin Reserve for field research. In 2014 Tim Doherty, for his PhD research on the spatial ecology of feral cats in this semi-arid environment, began an extensive monitoring program to uncover more information about their habitats and habits, which will be a vital link in the huge task of limiting the numbers of the most efficient killer of small native mammals, birds and reptiles.

Monitoring was essential at Edgbaston as Bush Heritage prepared its strategy for protecting the red-finned blue-eye. In 2008 Adam Kerezsy, a desert fish expert, visited the property to consider the protection project. He was initially disappointed when shown the spring where natural history enthusiast Peter Unmack had made his startling discovery, but quickly became enchanted with the goal of saving the fish from extinction. Because of the arrangements for the use of the family homestead that Bush Heritage had made with the Wills family, Kerezsy and other researchers camped on the property, usually on the clayplan close to the vulnerable springs.

Kerezsy's unique scientific qualifications for this job had been achieved after a chequered career which began with an Arts degree from Sydney University in 1989. An invitation to participate in an adventure sponsored by Dick Smith in the *Australian Geographic* changed his life. He and a friend, James Woodford, agreed to be the first white people to follow the Berkeley River in the Kimberly to the sea. They were dropped into the remote area by helicopter, and, with minimal supplies of food, travelled for two months. They saw no one, and experienced emotional and physical stress, particularly during the last week which they spent on a rock at the mouth of the river, waiting for the army to pick them up, with no idea when they would be rescued. While there, the two young men promised each other that they would strive to continue to lead interesting lives, and Kerezsy attributes the directions he has taken in later years to that moment. He had always been passionate about natural history, and completed a Graduate Diploma in Aquaculture. Living at Lake Cargelligo in central New South Wales, he began to consider fish in arid landscapes, and started a PhD on desert fish, based on several trips he had made to little-known desert water catchment areas. His book, *Desert Fishing Lessons: Adventures in Australia's Rivers*, tells the story of discovering fish in waterholes hundreds of kilometres from their usual habitats.[15] When he began work at Edgbaston, he was one of just a few acknowledged desert fish experts in Australia. He believed saving the red-finned blue-eye would probably be impossible, but quickly discovered that the most difficult aspect was navigating regulations, especially stringent because the snails and plants in the spring complex were also endangered. Since Edgbaston contains one of the few areas of artesian springs where the majority of its flora and fauna have remained intact for millenia, a complete 'microhabitat' had to be considered.

Ultimately, ensuring the red-finned blue-eye's survival meant ridding the springs of gambusia. Kerezsy and Rod Fensham, from the Queensland Herbarium, began physically removing the gambusia from one small spring, a flawed plan because it was unachievable in the bigger springs. The solution was to kill the gambusia in springs where the red-finned blue-eye was not present, with the chemical Rotenone, and then move the threatened fish into these cleansed springs. Rotenone had to be tested to determine the lowest effective dose, and to ensure that it would not harm the other species in the spring. Trialling different amounts of the chemical took time, and many submissions to government agencies. Once this was achieved, another year of applications for permission to spray the springs followed. On obtaining the permits, they sprayed two springs in winter, without adverse effects on the other inhabitants of the springs.

After two years, two springs were clear of gambusia, ready for the relocation of selected colonies of red-finned blue-eye. The fish lay one egg every three days, and live for several years, so that each school of fish comprises fish of varying sizes, albeit up to only 3 centimetres! Watching them swimming, predator-free, was the reward for the effort. By 2012 Kerezsy was cautiously confident that the strategy had been successful, and that it could be repeated in other springs.

Ecologist Sandy Gilmore has overseen ecological restoration on SA and NSW reserves.

Source: Bronwyn Willis

But gambusia could recolonise these clean springs after big rain events, when water flows across the land. After much trial and error, Kerezsy decided on low shadecloth fences dug into the ground to encircle the purged springs, which would not impede the natural flow of water when it came. Because of the harsh conditions, and the prevalence of feral cattle, kangaroos and emus, regular checking that the barriers remained intact was essential.

The other major threat to the red-finned blue-eye is one Bush Heritage cannot control. The springs are permanent sources of fresh water with a constant temperature. If the level of water in the Great Artesian Basin should fall too low, it will not pump its life-giving water into the springs. Free access to the water by agricultural and pastoral companies, by town populations, and by the mining operations that are springing up in the regions above the Great Artesian Basin, make this scenario probable in the future. As a safeguard, a captive breeding program has been developed to ensure the continuation of the species.

Other researchers also spend time in this bleak landscape. Renee Rossini, a PhD student from the University of Queensland, enjoys the subtle charm of the shallow springs while studying the nine or more species of tiny snails found here. Like the red-finned blue-eye, they have 'evolved both morphologically and physiologically ... so that they cannot live anywhere but these particular pools'.[16] Their ancestors saw 'the Australian interior shift from rainforest to an inland sea, to a vast dry and salty desert. And [they] have held on through every step of the way'.[17] Today, in each small pool there are 'hundreds of snails, shrimps and bugs crawling about the forest of weird and wonderful plants ... in dark, dirty pools, in trickling floral meadows and in the ephemeral "bad lands", to be discovered if one chooses a micro rather than a mega perspective'.[18] Renee understands the special nature of the springs. Because of their very particular flora and fauna, springs are usually hotspots of cultural and biological significance throughout the world, and Edgbaston is the 'most astounding example of this phenomenon in Australia', with at least fifteen species here found nowhere else in the world.[19]

In 2010 work on the Gondwana Link project increased, with more seed harvesting and planting on Bush Heritage properties. Angela Sanders, Bush Heritage's WA ecologist, was overseeing the process alongside her work at Kojonup, Charles Darwin and Eurardy reserves, until it became clear that Gondwana Link needed a dedicated manager.

New Zealander Simon Smale had worked as a landscape architect in conservation for over twenty years, first in the Department of Conservation, where creative design work took him all over New Zealand, and then as a conservation advocate in the Environment Court, as well as being the front man on the Maori television channel promoting conservation land management and gardening. He travelled extensively, working in Papua New Guinea,

Ecologist Angela Sanders and landscape manager Simon Smale, the forces behind the success of Bush Heritage's work in Gondwana Link

Source: Jessica Wyld Photography

Indonesia, Laos and Fiji, before a radio documentary about Gondwana Link caught his attention. In 2008 he took up a job with Greening Australia (GA) as program manager on the Gondwana Link project, where he worked for eleven months before becoming Bush Heritage's Gondwana Link manager.

The concept reignited his interest in integrating protected natural areas into the agricultural landscape, and in landscape restoration on a large scale. Smale revels in the variety of tasks: working on the land, buying properties, and explaining the restoration processes to Bush Heritage supporters. He also endorses Bush Heritage's commitment to 'start growing things and learning from it', rather than waiting for the grand plan to materialise. In this region of low rainfall, it takes time for the landscape to recover from damage. The mosaic of different soil types is key to understanding the variety of vegetation types, and Smale says it requires dedication and patience to recognise the fine gradations in the soil structure that produce different growth patterns and different plant species. While restoration through direct seeding and intensive planting is expensive, about $2000 per hectare, if balanced against establishing a landscape that will continue to evolve over hundreds of years, the cost seems negligible.

Sanders and Smale work together, Smale attending to the direct seeding, weeding and supplementary planting, while Sanders focuses on monitoring the ecological outcomes. They confer constantly, augmenting each other's knowledge and experience, their ultimate objective the creation of habitat for fauna which in turn will support healthy, robust plant life, and already there are fallen branches and leaf litter on the properties. Evidence of returning birds and small mammals animates Smale and Sanders, and they are excited by the number of birds flitting between the trees. Birds in Australia are the great pollinators, their symbiotic relationship with flowering Australian trees and shrubs creating a 'honey-eater pollination syndrome'.[20] Most flowering trees in Australia produce large quantities of carbohydrate as nectar in response to poor soils, and this attracts many species of birds. Australian birds are aggressive, and battles over nectar-filled flowers result in wide dissemination of the pollen. Because birds fly far in their search for flowers, pollination occurs at considerable distances from the source, ensuring robust genetic stock, and a continual source of new plants augmenting the direct-seeding program.

Sanders has chosen to manage Kojonup herself, and is in discussion with relevant authorities about incorporating the adjacent nature reserve into the Kojonup property. She is delighted with the success of the new technique she developed with Western Australia's Department of Parks and Wildlife to save the red-tailed phascogale from extinction. Between 2010 and 2012, they relocated thirty phascogales from the Mid West to Kojonup, placing them in wool-lined nesting boxes designed to protect the tiny tree-hopping marsupials from feral predators. The creatures build nests of sticks and leaves in the boxes and keep their young in them while they are still suckling. Bred on the reserve, they continue to use the boxes. Sanders has monitored their progress, reporting complete rejuvenation of the population: the threat that they will become extinct in this area has been lifted.

Smale revels in the variety of tasks: working on the land, buying properties, and explaining the restoration processes to Bush Heritage supporters.

Sanders shares her workload with Vanessa Westcott, who joined her as the regional ecologist for the WA Rangelands in 2013. Westcott is a botanist, and they integrate their research into larger ecological studies that connect the flora and fauna on each property. Westcott's first blog for Bush Heritage, written in March 2014, soon after her appointment, illustrates the range of work an ecologist undertakes. In her first months she set up sand pad monitoring at Charles Darwin Reserve, where fresh malleefowl tracks had been discovered, to everyone's delight, as it confirms that this area remains a stronghold of this threatened species; undertook stick-nest rat trapping with the Australian Wildlife Conservancy (AWC) team at Mount Gibson, the adjoining property; found a priority-listed plant species, *Androcalva bivilosa*, at Eurardy with an officer from the Department of Parks and Wildlife Threatened Species Group; went bat catching and pitfall trapping with WA Conservation Council's Nic Dunlop, and the citizen science team at Charles Darwin Reserve; and led a team of volunteers out to Eurardy to survey for threatened orchids. She also attended a Fire in the Rangelands conference in Perth to help with her upcoming work with the Central Desert Native Title Service in the Birriliburu Indigenous Protected Area (IPA); visited Len and Val Warren, dedicated Charles Darwin Reserve volunteers, to learn the history of Charles Darwin Reserve and Eurardy; visited Eurardy for a comprehensive exploration of the property; and met with community groups, Rangelands Natural Resource Management, Birdlife, the Geraldton Herbarium Group and Orchid Society. In later blogs Westcott reports more activities: she led fifty delegates from the International Association of Vegetation Science conference on a discovery trip to Eurardy; discovered native burrowing bees at Eurardy; worked with Ian Hamilton (Eurardy's reserve manager at the time), and amateur lepidopterists Mark Heath and Paul Kay, on surveying moths at Eurardy to contribute to the International Barcode for Life, an ambitious global collaboration of biodiversity students to create a 'DNA barcode reference library for all multi-cellular life', established by the University of Guelph in Canada.[21]

This crowded snapshot of Westcott's work illustrates the breadth of activities and in-depth knowledge that an ecologist must cover. It also highlights how Bush Heritage is working in mutually benefiting partnerships with local and state groups. The reciprocal interchange of scientific expertise, historical knowledge and on-the-ground shared experiences encourages further conservation activity and enthusiasm at many levels.

Changing the focus

By 2010 Bush Heritage had passed through the stage of intense property purchase that characterised Phillip Toyne's presidency. Toyne was the perfect president for the era of expansion and there is no doubt, that without his leadership, Bush Heritage would not be in the position it is today.

IN THE NINE YEARS he was President, and during which he enjoyed a productive partnership with Doug Humann, Bush Heritage went from owning twelve properties totalling 61 786 hectares to owning thirty-one properties with a total of 946 000 hectares, establishing Bush Heritage as a major player in Australia's conservation efforts. He oversaw a rise in income from $966 000 in 2000 to $17 843 000 in 2009, and initiated, and pushed through against stiff opposition, the relocation from Tasmania to the mainland, watching the organisation grow to over sixty employees, and transforming the Board into a more professional body with a strategic focus. He was proud of Bush Heritage's achievements and described it as 'far and away the most satisfying environmental engagement' of his extensive career.[22]

But times were changing. With a much less buoyant financial climate, there was a need to set a different focus for the future. The new President offered that possibility. Louise Sylvan became a Board member in 2007, but her involvement with Bush Heritage dates back to 1996, when she effected the transfer of her late husband's property, Burrin Burrin, from the Australian Conservation Foundation to Bush Heritage. She continued to support Bush Heritage, in 2007 donating her own property, 55 hectares of secluded bushland deep in a valley under the escarpment behind the NSW coastal town of Berry. 'Nameless' Sylvan Reserve is composed mostly of regrowth forest, old growth eucalyptus trees with broad canopies scattered among saplings that speak of a past when it was a temperate savannah, and a thick understorey of ferns, a dark and mysterious place where Richard Sylvan is buried. After Toyne asked her to join the Board, her elevation to President was by seamless agreement of all members following Toyne's retirement at the end of 2009.

Board members identified people whom they believed would bring the right skills, competencies and passion to the Board to enhance the organisation. The final three years of Toyne's term as President had been gruelling, raising hard questions about ongoing management and fundraising. That the Board was keen to explore different options indicated a growing maturity and a readiness to tackle change. In electing Sylvan, Board members were searching for a nuanced approach to conservation, one sensitive to factors that had radically changed the world since Brown first articulated his vision. Sylvan's background suggests a complexity and richness of understandings that range freely in a modern technological world.

She grew up in Canada, in prairie country. Her father was a prospector who possessed extraordinary bushman's skills. Accompanying him on his hunting trips, her eyes were opened to nature and the spiritual and pragmatic legacy of the indigenous communities from whom her father learnt his bushcraft. She studied business administration and commerce at the university in Alberta, Canada, followed later by a Liberal Arts degree and graduate work for an MA in public administration and policy in British Columbia.

Marrying Richard Routley (who later adopted the surname Sylvan) brought her closer to the environmental world and she joined her husband in philosophical discussions on the ethics of care for the natural world. After emigrating to Australia, she engaged with the theoretical aspects of management and public policy, in 1987 becoming the inaugural Executive Director of the Consumers Health Forum, a national health consumer organisation set up by a group of NGOs with financial assistance from the federal government. Becoming a strong business-oriented advocate, with involvement

Bob Brown and Paul Thomas gift Oura Oura to Bush Heritage. From left: Paul Thomas, Bob Brown, Bush Heritage president Louise Sylvan and Doug Humann

Photo: Peter Morris

in policy debates on financing and regulation of healthcare, she helped change the dialogue from one dominated by the health providers to one that spoke for the consumer, becoming the Chief Executive of the Australian Consumers Association in the mid-1990s.

Sylvan is comfortable challenging powerful vested interests, and unfazed by resistance to her ideas or personal attacks. She tends to create forums in which dissent and discussion breed better solutions. In 2003 federal Treasurer, Peter Costello, invited her to stand as deputy chair of the Australian Competition and Consumer Commission. Five years later, she was appointed as a commissioner with the Productivity Commission, and participated in a number of major enquiries.

Sylvan's experience in advocacy and public policy has meant that Bush Heritage has at its helm a woman practised in complex and sophisticated thinking and used to dealing with change. Toyne had recognised her capabilities, and that her heart was engaged as well. Not a person to take the limelight herself, she believes that Bush Heritage is best served if the CEO stands as the embodiment of the organisation. Previous presidents had also endorsed this model, but under Sylvan the role was clarified, as was the Board's role of scrutiny, support and long-term strategy development.

Sylvan began her presidency at a time of unrest at Bush Heritage. The focus had been on property purchase, and perhaps the expansion phase had happened too quickly and too broadly, when more money and effort should have gone into increasing staffing on properties, the subtext being that there was always money to buy properties but not necessarily to manage them, a tension that had always existed. Ecologists were thinly stretched, often travelling enormous distances between the properties that they looked after, and working on projects that might never be developed, because there was not enough money to do everything, nor enough staff to undertake the work.

The Board sought ways to ameliorate the situation. The Marketing and Fundraising departments focused on activity on the properties in their communications with supporters, gradually steering away from stories about purchases to the work of conservation. And Annette Stewart as systems specialist was showing how remodelling the underlying systems on which everyone depended could be beneficial.

Sylvan is comfortable challenging powerful vested interests, and unfazed by resistance to her ideas or personal attacks.

Ecologists in the Conservation unit were asked to develop a method to define the organisation's mission: 'to conserve land, water and wildlife of high conservation value' by biological and ecological productivity measures within the Anchors in the Landscape framework.[23] The project, known by its acronym 'BioPrEP' (Biodiversity Prediction using Ecological Processes), was driven by conservation manager Paul Foreman and written with ecologists Sandy Gilmore and Murray Haseler, in partnership with ANU's commercial arm, ANU Enterprise Ltd, and Hugh Pringle, now a private ecological consultant who was then working for Bush Heritage. A highly technical document, it draws upon data from scientific, environmental and conservation agencies throughout Australia, making it a comprehensive compilation of the current state of ecological knowledge that supports the scientific work of the ecologists. Investment decisions in the future would be informed by the explicit and systematic set of ecological considerations set out in the BioPrEP document of 2010, and could be fed into Bush Heritage's own custom-built organisational system that Stewart was building.

In 2011, after fourteen years at the helm, Doug Humann decided to retire as CEO. The job had changed, and he believed he had taken the organisation as far as he could. Predating Toyne's arrival as President by several years, his work had prepared the ground for what became a hugely successful partnership between the two, and together they had transformed Bush Heritage. Two years after Toyne's departure, Bush Heritage had continued to grow, achieving more than seventy staff in twelve locations, and further properties and partnerships. During Humann's time as CEO, Bush Heritage had become respected within the conservation, not-for-profit, business and government sectors, had developed a successful fundraising program, and had gained a seat at many decision-making tables. Looking back, Humann recalls:

> There was so much change at Bush Heritage. We were constantly growing, overcoming challenges and inventing new ways of doing things. There was never a chance of getting bored. Leaving Bush Heritage was one of the most difficult decisions I have made professionally, and I'm grateful that the Board left the timing in my hands, and gave me great support and encouragement.

Humann had worked hard, investing himself in building the organisation. He had travelled extensively, developed connections with influential organisations and powerful people, and helped build Bush Heritage's reputation as a collaborative and innovative organisation. As did Toyne, he has continued his interests in conservation through a consultancy role, encouraging Aboriginal groups to join the Caring for Country movement, and assisting them as they find their way through complex bureaucracy to achieve ownership of their lands. In 2014 he received the Australia Day honour of Member of the Order of Australia (AM) for significant service to conservation and the environment.

Some in the organisation felt Humann stayed a year too long, on top of an illustrious career that had set Bush Heritage on course. And Humann himself, with his straightforward honesty, agrees.[24] A new CEO would bring different energy, and effect necessary changes in policy and staff activity.

Gerard O'Neill was travelling with friends along the Anne Beadell Highway, a red sandy track winding through the Great Victoria Desert National Park, connecting Coober Pedy in South Australia to Laverton in Western Australia, when he received the call for a second interview at Bush Heritage for the position of Chief Executive. He was driving the lead car, with the window down, and it was a slow journey. His passion for birds and his keen ear meant that the group stopped frequently for O'Neill to leap out of the car, binoculars in hand, to follow a faint bird call or identify a fluttering wing glimpsed through the trees.

O'Neill's past experiences make clear why Bush Heritage was keen to employ him. Growing up in the southern Riverina, knowing 'things about the land, the weather and climates' and about identifying different kinds of birds, he began work as a stock and station agent and developed an understanding of cattle and horses while travelling in rural Australia, honing his skills. But the job was not intellectually challenging, and he decided to study Natural Resource Management in Canberra. People thought he was mad, leaving a good job with career opportunities, but he knew that he had found his life work. It was 1977, and he was in the second intake of a new strand of science (Karen Alexander was in the year ahead), at a time when the implications of resource management were being defined academically and legally. After graduating, O'Neill was in the first group of rangers to work in Kakadu National Park. Policy changes were rife, and emotions ran high with the volatile mix of land rights, uranium mining, ecological arguments and radical left-wing politics. Under the scrutiny of the press, several distinguished, intellectually stimulating and well-resourced leaders explored the dynamic of land rights and land use conflicts in a national park.

The young O'Neill made his own judgments. He rejected the idea of 'wilderness', a philosophical position at the heart of much modern conservation work that loses credibility when considered against the historical manipulation of the landscape. His encounters with Aboriginal people made him realise that wilderness was a Western concept and in Kakadu he was working with 'natural areas' and 'cultural landscapes', created and nurtured by Aboriginal people over time. He developed his skills in reading the country, with a view to managing and conserving it, receptive to the thinking of the Aboriginal people he met.

He had known about Bush Heritage since the 1990s, and thought the campaign slogan, 'We don't beat around the bush, we buy it', quite 'cheeky'.[25] He knew Doug Humann and admired his handling of the cattle in the high country campaign. When Humann moved to Bush Heritage, O'Neill began thinking about his own career. In 1995 he became an executive in the then Victorian Department of Conservation and Natural Resources before transferring to Parks Victoria in 2001. In 2007 he was appointed deputy chief of Parks Victoria, taking on the additional role of chief operating officer. After Humann resigned from Bush Heritage, O'Neill was considering applying for the position when he was approached by Bush Heritage to apply.

The culture of success gives people the freedom to have a go, make mistakes, and innovate, while understanding that they are supported.

In a competitive field, O'Neill's experience impressed the Board. He had worked in the tropics, in arid and in temperate zones, from the ground up, from cleaning toilets to advising government ministers on policy, and negotiating major deals, managing big budgets for organisations with over 1000 people. But the qualities that shone through were his strategic thinking and the instinct to innovate. In 2015 these qualities are still evident, as he juggles day-to-day office issues with plans for developing partnerships, and confidently handles public sessions at Bush Heritage presentations. He can weave unexpected questions into the complex journey that Bush Heritage is undertaking without losing focus on the main message. He exudes the suppressed excitement and energy of a man with much to do, which contradicts, while still enhancing, the close attention he gives to the matter in hand.

O'Neill and Sylvan have brought different approaches to strategy to the table, but agree on basic principles. They know the battle for the environment can only be won if it is waged equally on the ground and at the policy level. Sylvan believes that, as a landholder, Bush Heritage is part of many communities, and she has urged reserve managers to play pivotal roles, wherever possible, in the communities in which they live and work.

To do this succcessfully, O'Neill believes that reserve managers have to have confidence in their decision-making. Under Humann, administration was centralised in Melbourne, with a traditional hierarchical structure. O'Neill has overturned this, giving reserve managers the authority to innovate, to make decisions for which they are accountable, changing their relationship to their reserves and their interactions with local communities. O'Neill sees himself as a 'transitory being' in the continuing story of Bush Heritage, and is developing a leadership culture which he hopes will remain long after his departure. The culture of success gives people the freedom to have a go, make mistakes, and innovate, while understanding that they are supported. There is room, O'Neill says, to take 'good-quality risks', based on educated reasoning to achieve positive environmental outcomes.[26] Sylvan and the Board back O'Neill's approach, believing that if the right business structure is in place, the decisions that individuals make will flow from a solid foundation.

While O'Neill was advocating these broad principles, Annette Stewart's new management system was being rolled out. Once accepted, it gave reserve managers the tools with which to assert their authority, and made it easier for them to take control of their areas of expertise and responsibility. At the same time, Bush Heritage, guided by former staff member Stuart Cowell, found another use for the Open Standards for the Practice of Conservation. Cowell and others adapted the methodology so that it is culturally appropriate and relevant to Aboriginal people looking for a framework around which to manage their country. 'Healthy Country Planning' is a key component in establishing relationships with Aboriginal groups, and has been the basis for the effective partnerships that have developed.

Aboriginal partnerships

In 2016, Bush Heritage has established, and is continuing to explore, partnerships with a number of Aboriginal groups. From the start Phillip Toyne, and Doug Humann had been keen to develop these partnerships, and did not neglect this aspect in the mid-2000s while pursuing traditional land purchases.

THEY ENCOURAGED STUART COWELL and Sarah Eccles, who were managing the partnership project, as they met with Aboriginal groups in each state to explain how Bush Heritage could support them in developing their own plans to care for their country. While each situation offered different opportunities, and individual approaches, the steps taken by Bush Heritage were consistent, since their aim was to develop trust in, and acceptance of, the reciprocal value of 'two-way knowledge', that would underpin conservation of fauna and flora, and ensure better outcomes for the people involved. In many cases collaboration on a Healthy Country Plan was, and will continue to be, the starting point for an evolving partnership in which trust and friendship are important components. The story of the evolution of the first Healthy Country Plan, which occurred as a result of Bush Heritage's collaboration with the Wunambal Gaambera in the far north Kimberley region in Western Australia, illustrates the point.

The Wunambal Gaambera people had registered their native title claim in 1999, and were already working towards the time when their land would be declared an IPA. Mutual trust and understanding between Cowell and the Wunambal Gaambera grew over several years, as the two organisations collaborated, using Healthy Country Planning as a guideline, with the objective of ensuring that country would remain 'healthy' (in Aboriginal terms) and 'protected' (in Bush Heritage terms). Wunambal Gaambera country includes 925 000 hectares of freshwater land of eucalypt woodland with deep green, forested riverbanks and rainforest zones, and 1.6 million hectares of sea country with over 1000 islands in the north-west Kimberley region in Western Australia. It is identified nationally and internationally as a relatively stable and intact priority eco-region for global conservation, with numerous endemic fish, reptile, bird and mammal species, many of them rare or threatened, and with several vegetation and ecosystems at risk. Its rich overlay of complex cultural landscapes housing rock art from both *Wanjina* and *Wunggurr* (creator spirits) and Gwion (Bradshaw) beings add to the special identity of the area. Like all groups who choose to work with Bush Heritage, the Wunambal Gaambera harbour strong connections to their land, and want to live on country to maintain their spiritual and societal values. They are luckier than most groups since they already have some agreements with pearling companies in their waters, and the Mungulalu Truscott Airbase from which charter companies operate fly-in fly-out services to offshore oil and gas rigs. In meetings with small groups, communities established a priority list of key projects that they wanted to pursue, pinpointing the threats to the successful prosecution of those projects. The Healthy Country Plan gained general community acceptance because it was a synthesis of the ideas and aspirations of the Wunambal Gaambera diaspora living in and around Kalumburu and Derby and further afield, with whom Cowell participated in field trips for research and land management, and joint 'on country' planning workshops. They came together to amend and endorse the draft plan, and the final document, put together by Heather Moorcroft, conservation planner and geographer, is visually stunning. While it employs simple language that is accessible to all its readers, it is also elegantly descriptive, with a wealth of detail, and is invaluable as a model for later Healthy

Wunambal Gaambera partners manage over 2.6 million hectares of land and sea country.

Source: Wunambal Gaambera Aboriginal Corporation

Bush Heritage Chief Executive Gerard O'Neill (left) and board member Chris Grubb (behind) signing a Memorandum of Understanding with Birriliburu Traditional Owner Ben Brown and Rob Thomas from Central Desert Land and Community

Source: Gerard O'Neill

Country Plans. Once the IPA was established in 2011, Bush Heritage's support in bringing the Healthy Country Plan to fruition was both critical and strategic.

An issue for some Aboriginal communities setting up corporations that will manage land is that, although the federal government will pay ranger wages, put fuel in vehicles and purchase capital items, there is little support for the corporations to develop internal capacity and become sustainable.[27] To solve this, the Wunambal Gaambera Board employed an operations/business manager, and Bush Heritage embedded the paid position of healthy country manager within the Wunambal Gaambera Aboriginal Corporation (WGAC) to support the acceptance, implementation and reporting on the Healthy Country Plan. Tom Vigilante began work in 2011, keen to continue Bush Heritage's commitment to 'two-way knowledge' as an active ingredient of Bush Heritage's relationship with the Traditional Owners.

Vigilante, no stranger to the North Kimberley area, initially became interested in Aboriginal land management when he visited his grandfather, a long-time volunteer market gardener in the Kalumburu community. He completed a PhD on the ethno-ecology of Aboriginal burning around Kalumburu, and worked for the Kimberley Land Council for eight years on Aboriginal natural and cultural resource management projects, including fire management. When Bush Heritage appointed him, he was accepted into the community because he was already known and because his role is supportive, rather than managerial. The Traditional Owners, board members of the Wunambal Gaambera Aboriginal Corporation, instigate and control all activities, and make decisions encompassing cultural concerns which all employees respect. Vigilante works with Neil Waina, the head ranger, and his team of six Uunguu Rangers and numerous casual Traditional Owner rangers, to help organise logistics, oversee Healthy Country Plan implementation, and develop strategies to support the plan. More recently WGAC established the first ranger station at Garmbemirri, not far from the Mungalalu Truscott airbase and close to the coast, which facilitates expeditions to care for and monitor the health of both land and sea.

Houses and other facilities are currently being built at Garmbemirri, the first of several homelands for those wishing to leave Kalumburu to live on the land that nourishes them spiritually and physically. The current chair of the board, Catherine Goonak, already lives with her family at Kandiwal on the Mitchell Plateau, and, as the next step after the completion of Garmbemirri, Wunambal Gaambera plans for this small community to grow, with other community homelands to follow. The Healthy Country Plan has a ten-year life, from 2010 to 2020. At the end of this time an evaluation of the goals achieved can be made, and a new plan developed if required. A progress review by an outside panel has indicated that the group, now halfway through the plan, is already ahead of its goals in its Healthy Country Plan.

The Traditional Owners, board members of the Wunambal Gaambera Aboriginal Corporation, instigate and control all activities, and make decisions encompassing cultural concerns which all employees respect.

Bush Heritage's contribution is appreciated for its non-interventionist, non-managerial approach. While Vigilante is a Bush Heritage employee he is comfortable working across the two organisations and managing the interface between the cultural landscape of the Wunambal Gaambera people and Bush Heritage processes. Working with Bush Heritage also gives him access to the networks that can assist the community in decision-making. For Bush Heritage, there is pleasure in investing in a partnership that enables other groups to carry out programs to protect Australia's plants and animals.

Another partnership model has developed with the Martu people, Traditional Owners of Birriliburu country in the Western Desert region in Western Australia. They achieved native title to their land in 2008, quickly establishing an indigenous ranger program, in which the rangers use their traditional, ecological and cultural knowledge to preserve and protect plants, animals and their culture in the region, leading teams to restore traditional fire practices, monitor threatened species and conduct baseline fauna surveys. After their IPA of over 6.6 million hectares of desert country (an area roughly the size of Tasmania) was established in 2013, they welcomed a partnership with Bush Heritage, and the two groups have shared their excitement and pleasure in watching the country renew itself. The IPA covers three bio-geographic regions: the Little Sandy Desert, the Gibson Desert and Gascoyne, a diverse and beautiful country of sand dunes, sandstone mountain ranges, salt lakes and claypans, and home to nationally significant species such as the bustard, bilby, mulgara and marsupial mole.

Ecologist Vanessa Westcott works with the Martu and the Central Desert Land and Community Team, learning much from their profound knowledge of the land and its fauna and flora. In 2015 the partnership was formalised for another three years, with Bush Heritage providing funds for equipment, project resourcing, ranger wages and the continuing work of Westcott. Chief Executive O'Neill, on a visit to confirm the agreement, was impressed with what the Martu have already done to develop the ranger program. The partnership has flourished, bringing gains in knowledge and shared motivation for everyone. For Westcott and other Bush Heritage participants, watching the Martu read their country and its ecology is a privileged learning experience. For the Martu, particularly the women, who have given their time and passion to conservation, there have been gains in personal health and self-pride. The local health service believes their involvement has added ten years to their life expectancy.

Garawa man Jack Green, still waiting to establish an Indigenous Protected Area at Robinson River, NT

Photo: Peter Martin

In North Queensland, Emma Ignjic, Aboriginal partnerships manager, is project managing existing partnerships, and developing new and emerging ones. Signing an MOU, however, is only the first step, and unless the people on the ground have the capacity and invest sufficient time, the partnership will remain hollow. As more Aboriginal groups turn to Bush Heritage for support in developing plans for the care of their country, Phillip Toyne's vision, which in fact had less to do with ownership of the land and the number of hectares saved, and more to do with the quality of the conservation experience, becomes more apparent. One-third of Australia's ecology resides in lands for which Aboriginal people have ownership in law. As the methodology to measure the progress of individual programs is the same, whoever owns the land, Bush Heritage believes that where knowledge and objectives overlap, and where different stakeholders can walk together across the land, the possibilities for achieving successful conservation outcomes are greater, and will be evident over ever-broader landscapes. While in some cases Bush Heritage has been able to promote its management influence in these lands and, with agreement from its Aboriginal partners, include these areas in the tally to save 7.7 million hectares by 2025, the emphasis is now on remaining relevant, influential, and connected to the many different types of partnerships that can achieve large-scale outcomes. Bush Heritage knows how fragile these relationships are, that progress in partnerships can alter with a change of leadership within the Aboriginal community, and that partnerships must be seen in the context of its avowed intention to be a conservation agent for the long term. Such setbacks mean that it must wait impartially, treading lightly, and exercising patience. But there is always a risk that projected programs and activities out-pace the people on the ground, as Jack Green's story shows.

Green, senior Garawa man and a Traditional Owner of country in the south-west Gulf of Carpentaria, has worked for the Northern Lands Council. He is well versed in the politics of land claims and government regulations, having been involved in efforts to reclaim land since the 1970s. In 1984 the Waanyi Garawa Aboriginal Land Trust won a grant to buy the pastoral lease of Nicholson River Station, and in 1990 the Garawa Land Trust obtained funding to buy the pastoral lease of Robinson River, of which Green is a Traditional Owner. The efforts of both the Garawa and Waanyi peoples to return to live on their land has been fraught with setbacks, as have their aspirations to combat invasive weeds, and thus reduce the wildfire risk. They have watched massive fires tear through their land, and have resolved to begin Caring for Country projects to combat their greatest environmental problem: fires burning too strongly, and too frequently. The federal government's Working on Country program brought many Traditional Owners back to their land to spend time on country, all the while transferring their stories and skills to the young people. Since beginning work with no resources, no equipment, and much self-doubt, they have seen gradual improvements on their land. Changing the fire patterns has encouraged more 'bush tucker', and among the rangers and others involved in the programs there is a palpable growth in pride and self-confidence through achieving progress under difficult circumstances.[28]

Ignjic and Aboriginal partnerships manager Justin McCaul helped set up the IPA planning meeting held on country, in itself a massive logistical undertaking to bring everyone together from far-flung points. Despite outside disruptions from WA mining companies attempting to influence outcomes by offering payments to Traditional Owners to consider mining opportunities on their land, and the intrusive actions of neighbouring pastoralists offering free meat for the chance to negotiate on pastoral leases, many issues were resolved in favour of working with a Healthy Country Plan.

But in 2016 Green is still waiting for his application for an IPA on the Robinson River area to be granted. He knows that the success of this venture depends not so much on the money as on the on-ground support of people with the skills and expertise to guide the group through the make-or-break governance issues. He is supported by Seán Kerins, a research fellow at the Centre for Aboriginal Economic Policy Research in Canberra, and both believe that Bush Heritage is best suited to deliver the structured, effective help needed at this point. Bush Heritage is also keen to work with Green and Waanyi Garawa head ranger, Jimmy Morrison. But in the time between McCaul's departure in 2012, and ecologist Allana Brown's arrival in 2014, Ignjic has worked solo. She and Green have been frustrated in their attempts to meet again, due to work commitments on both sides and the fourteen-hour drive from either Cairns or Darwin to get to this remote corner of Australia. It is easy to conclude that Bush Heritage personnel are spread too thinly, but both sides hope that the partnership will progress in the near future.

Traditional Owners implementing 'right-way fire' on Birriliburu country

Source: Dave Whitelaw

Caring for country

Coming from Ramingining, in north-east Arnhem Land, we bump along the top road towards Oenpelli over severe corrugations and numerous river crossings fringed with pandanus and paperbarks. We are on our way to the Warddeken ranger station at Kabulwarnamyo will meet someone from the community at the Maningrida turn-off. The donor who wanted to fund the Wollogorang purchase was encouraged to support employment arrangements on the Warddeken ranger station and I am keen to understand what this has meant.

A week ago, we were on the Savannah Way, heading to Borooloola, and passed the turn-off to Wollogorang. We had just learnt that Wollogorang had been sold to a Chinese company, no doubt interested in its potential as a cattle station. Seeing the sign, I felt, once again, sharp regret for the station that could have been one of Australia's most significant conservation reserves.

Jake Weigl, the land management coordinator for Wardekken, and Dean Yibarbuk, a Traditional Owner and resident fire ecologist, meet us at the turn-off. So far, so good; we begin what turns out to be a four-hour drive south over more rough roads. We stop to cut the hind leg and backstrap off a buffalo that Jake shot on his way up to meet us – good tucker for the community.

At last we arrive at Kabulwarnamyo. And it is a revelation. Houses beautifully designed for the climate that cost $15 000 to build, with wooden slat platforms raised off the ground on steel foundations on which stand a variety of innovative 'fitouts', mostly insect-proof tent rooms for easy living. Cypress pine and steel poles support huge weatherproof tarpaulins that shade the whole, but let in the breeze. There is a school built along the same lines, still bare of furniture. The teacher shows us what she has planned, and where she will welcome her first students next week. It is the beginning of a bold innovation by the Warddeken Board for a community that believes in shaping its own destiny. Designed so that rangers can stay on country with their young families, it will rely on philanthropic funding. As we tour the site, we learn that people are focused on work and commitment to the community ideals, with a waiting list for people wanting to live here once more houses are built, even though it is still 'just' a ranger station, and very remote. We meet several of the rangers, some of whom were here at the beginning, over ten years ago, having chosen to live as much as they can off the land and teach their children the stories that will sustain them as they grow up proud of their heritage, and independent of town ways. We visit Mary, the widow of Bardayal Lofty Nadjamerrek, the charismatic leader who pushed to establish this ranger station, so that his country could be cared for in the traditional way, and families could come back and live on country, and be healed. His paintings from a lifetime on country, depicting the creation spirits and the fish and animals he hunted, are scattered throughout Warddeken land, on sheltered rock overhangs, on cave walls where the stone floors are shiny with thousands of years of living. In the afternoon Jake, Dean and Ray, Bardayal's grandson, take us on a tour of the country, and we visit the 'cool cave'. We feel the cold blast of air from 20 metres away despite the 30 degree day, and gasp in awe at the huge vertical, narrow fissure in the rocks that opens to an icy cave. Here Ray introduces us to his ancestors as he seeks their permission to show us this special place. Nearby is Bardayal's last painting, a bold white kangaroo on the rock wall, and the much smaller figure of a hunter with spear raised. Jake and Dean cut out a sugar bag from a eucalypt, and we lick the sweet liquid that drips as it is transferred into a clean bucket. We continue in the car, stopping beside the river, where Jake dons his climbing gear to reach another sugar bag that Ray and Dean, sharp-eyed and keen, have identified from a small raised bump with a black spot, high up on the smooth limb of a big eucalypt. The wild bees have long gone, but the sugar bags – there are two of them – are a good size and greatly prized, once half the tree has been cut down and they are eventually revealed. They will be picked over for the best bits by the old ladies when we get back. While the chainsaw brings down the branches, Peter and I dip into the crystal clear water of the Liverpool River, since the tree is conveniently located beside a gorgeous deep pool ringed by venerable melaleuca and pandanus, with purple flowers in wet mossy crevices on the rocks, a magical place of such serenity and antiquity that we break the silence only with the splashing of water.

Back at the ranger station, we are keen to understand what makes this community tick. Jake, doing 'the only job I have ever wanted', and his wife Georgia Vallance, operations manager, are central to activity, respecting and working with the Traditional Owners whose decisions guide the programs. Jake has been here nine years, Georgia seven years, and the continuity they provide is the glue that binds. The place is run by the board of Traditional Owners as an incorporated company limited by guarantee, signifying a break from total reliance on government funding for operational funds, and demonstrating Warddeken's desire for financial self-determination. The dream of coming home to Warddeken land was achieved through the funding stream derived from the successful West Arnhem Land Fire Abatement project, where five independent ranger groups undertake fire management of over 28 000 square kilometres of West and Central Arnhem Land. Funding and expert legal support from the WWF and Bush Heritage enabled Warddeken Land Management Limited to obtain charitable status through the taxation office, and to prepare an IPA Management Plan for 2009–13, a prerequisite for government recognition. In Darwin, CEO Shaun Ansell deals ably with the financial, legal and bureaucratic issues, submitting appropriate grant applications to expand the ranger program, and to enable weed control, fire abatement, and feral animal control programs. Ansell's position is paid for by Bush Heritage, and he acknowledges, as do Jake and Georgia, the strategic role Bush Heritage has played in supporting Warddeken people. 'Bush Heritage is by no means our only partnership, and certainly not our biggest funder, but their support has been crucial – they have funded the thing that no one else does, that is, support for personnel in the positions that provide good governance for the community.'[29] Once again, it is that subtle touch, the soft hand on the lever, that enables rather than manages.

We spend two nights at Kabulwarnamyo, our tent pitched beside a clear, quietly bubbling spring, and we bathe in the morning as steam rises between pandanus fronds and peeling paperbarks, around us the strange chuckle of the northern kookaburras with bright blue wings. The community is lucky: with two springs to bathe in every day, the children are free of the ear and nose problems that beset many young Aboriginal children who do not have the luxury of running water close by. Georgia tells me that Dr Paul Burgess, a 2014–15 Commonwealth Fund Harkness Fellow in Health Care Policy and Practice, considers this community the healthiest in Arnhem Land, and she is in the process of detailing the way of life that makes this community a standout as far as social return on investments go.

But this is only the start of our adventure on Aboriginal homelands. We spend a week camping with Warddeken rangers and families at Kudjekbinj, half a day's drive away, as they conduct a fauna survey on land that has never been surveyed before. The Traditional Owners of this country look forward to the time when they can start their own ranger base here and work with agencies to keep this country healthy, too. No one is under any illusion that it is going to be easy, but they speak of being on their land with a longing and a powerful joy that carries you with them. They envisage a time, like the Wunambal Gaambera people, when the land will once again be home to many communities dotted across their country, and each group will look after its particular region. Allana Brown, Bush Heritage's northern ecologist and healthy country manager, has joined Terry Mahney, an ecologist from Darwin, to help with the survey, and we spend a busy week, inspecting the six sites twice a day, opening the traps in the morning, collecting the lizards and small marsupials for identification, disposing of the bait and closing the traps. In the evening we return to rebait the traps, bringing back the animals we found in the morning to release back into the wild. Actually Peter and I do none of this; we are designated photographers for the groups. It is the rangers and their families, some of the children quite young, who do the work, guided by Terry and Allana, so that they can carry out the next survey themselves. It is quite hard, continuous work, but not one child complains, and all participate for the five days, threading their way through open grasslands, over rocky escarpments, and in thick semi-tropical vegetation that is green and lush against the cloudless blue sky. Two big rockpools beckon at the end of a hot walk at one survey site and I join the children in the water, daubing myself with the white viscous clay that they dig out from the bank, and we dive deep to wash it off. At night we go spotlighting, without much success, but enjoy the camaraderie of the group who share their knowledge of night critters with Terry and Allana.

I spend a lazy last afternoon with Terry, and Andrew, a Traditional Owner, and his children and several others from the group, on the riverbank where hot sunlight comes dappled through pandanus, palm and melaleuca, cool and green. The children reel in black bream with casual efficiency, and then Terry and I are lured into the river for an energetic game of chasey where we are outswum and outchased to exhaustion. As I dry off, Lizzie tries in vain to teach me how to peel back the skin on the pandanus frond to tease out the thread for weaving, a skill I can't master despite her efforts, and I notice her surprise at my ineptness. Andrew tends the fire in the pit he has dug to cook the fish, laying them on the glowing coals topped with fresh young melaleuca leaves. He covers all with a thick sheet of paperbark, packing earth around the edges to stop the steam seeping out,

Author Sarah Martin in the fun of a swim and clay daubing at Kabulwarnamyo

Photo: Allana Brown

and we sit around the fire, waiting for the fish to cook, as Andrew talks about his country, how it was, and how it will be. Delicious smells waft on the air, and I am filled with contentment, in love with this beautiful landscape that is full of contrasts, where red jagged escarpments enfolding secret subtropical glades with water running cool and deep give way to harsh, dry stretches of grassy woodlands, dry stony riverbeds and twisted jumbles of boulders tossed about like giants' playthings, an environment so absorbing, so rich and changeable, that it crowds out all other thoughts, demanding one's full engagement. It will be hard to leave and return to urban mundanity.

Much further south, in Western Australia, we have another, completely different experience of Bush Heritage's influence on the landscape. As visitors to the second Blues for the Bush festival held in October 2014 on Charles Darwin Reserve, we spend a friendly couple of days helping with preparations for an influx of nearly a thousand visitors. 'Locals' from Geraldton are setting up the staging and shady canopies and toilets. I join a group repainting huge movable papier mâché characters retrieved from Sydney's Gay Pride parade, which must be rendered more 'environmental', and Peter works with Fiona to create 'Mr Paterson' out of pieces of junk metal, of which there is a vast choice, and he is finally clothed in a vest and trousers made of rolled chicken wire, in which festival-goers will stick the twenty Paterson's curse ('Salvation Jane') weeds they have pulled, roots and all, from the earth to qualify for the raffle.

The morning of the festival dawns brilliant and clear, and it will be hot soon. We spend the day listening to talks by representatives of groups involved in the restoration and conservation of this semi-arid region who have joined the Gundawa Regional Conservation Association Project, an organisation which Healthy Landscape Manager Luke Bayley chairs. The mining industry's contribution of $2 million to offset environmental damage has benefited the Gundawa Association. Investment in climate change issues has brought the community together to think strategically about how to work more cooperatively to deal with the problems that are becoming more apparent. Presentations are inspiring, and the question-and-answer time draws audiences into useful dialogue and recommendations. As afternoon becomes balmy evening, the music and circus entertainment begin. With the setting of the sun, red and gold across the wide desert sky, the tempo increases, and a variety of performances keeps the audience transfixed or dancing to a vibrant beat under a star-filled sky till midnight.

Reserve managers

With O'Neill as Chief Executive, the office and land management cultures have changed. A new level of regional managers (later retitled 'executive managers') has taken pressure off the Melbourne office, and strengthened the line of responsibility back to the reserve managers.

THE COMPLEXITY OF THEIR JOBS has been addressed, too, by introducing an intermediary level of regional management with the creation of healthy landscape managers. While the executive manager is responsible for all the reserves and partnerships in a region, the healthy landscape manager concentrates on developing partnerships with communities, including Traditional Owner groups, in the specific Anchors in the Landscape areas now called 'priority landscapes'. The role of healthy landscape manager also includes oversight, or occasionally, direct management of the reserves within that landscape, to encourage a more significant role for Bush Heritage in the region, something that is especially important in remote areas. The position of healthy landscape manager is a fluid one, open to those with initiative, and the ability and desire to connect with groups within their communities. Reserve managers can aspire to become healthy landscape managers with a broader responsibility.[30]

On the whole, reserve managers relish their expanded roles, and enjoy the new relationship with O'Neill, who says, 'I've got confidence in you; run hard with your idea, and tell me how you are going when you are ready'. Employees in the field and in the national office know they can draw on his talents, his experience and friendship. They can use him for guidance, as a sounding board, or as a final port of call, and accept the matrix he draws: 'Here's where I want you to get to, and I think you are here, and I think you need (a bit more courage, a bit more training, or skill, or support, or experience)'.[31] He believes that 'collectively, we'll get more done rather than waiting for me to make a decision. We are all trying to move out of a centrist space into a more diversified leadership structure'.[32]

Glen Norris exemplifies this approach. At Bon Bon he had developed good relations with his neighbours, the mailman, the local Natural Resource Management team, contractors and suppliers. But after recovering from his injury on the Stuart Highway, he looked for another role. Keen to remain at Bush Heritage, and with the desert in his soul, he jumped at the opportunity to take up the healthy landscape manager position for SA to be based

at Boolcoomatta Reserve, purchased in 2006. He approached the new work with the confidence born of good experiences at Bon Bon.

He participated in the social life of staff of the local Natural Resource Management team, joined fire and fitness training, and reached out to neighbouring sheep farmers, sharing their anxiety about feral animals and weeds. Earlier managers had enlisted volunteers to refurbish the shearers' quarters, and the residences clustered near the homestead are now suitable for visitors, researchers and volunteers. He built closer relationships with the Adelaide universities, and there are plans to use the historic bluestone woolshed for artist and writer retreats. Norris also took on responsibility for local fundraising for Boolcoomatta, initiating media coverage for the property, and developing opportunities for the SA community to visit and become involved in conservation work.

Norris continued the work he had begun at Bon Bon with Traditional Owners. At Boolcoomatta, the Adnyamathanha people are still closely linked to the land, which has deep significance for them: 'Whatever happens to the land happens to us ... All these things have great cultural and spiritual importance for us, and we have stories about [the animals and plants and the rocks] that help us to understand them and keep them healthy'.[33] Long before Bush Heritage bought Boolcoomatta, the Nantawarrina rangers and Nepabunna community had been managing their country under Australia's first IPA arrangement. The partnership formed between Bush Heritage staff and the Nantawarrina rangers is furthering the conservation work on their country. In 2013 the rangers reviewed their management plan, working with Bush Heritage and other organisations to develop the plan for the next ten years, including a monitoring program for their conservation targets.

To the west, the Bimbowrie Conservation Reserve protects the yellow-footed rock wallaby that once ranged across the region where low rock outcrops provide good habitat. Last seen on Boolcoomatta in 1924, the wallaby was sighted by the previous reserve manager, Peter Ashton, in 2011, and Norris hopes the improved conditions on Boolcoomatta will see it return in numbers.

Norris suggested a joint partnership that might include the local organic beef business and BHP Billiton, whose work in the region supports good conservation outcomes.

Arid Recovery (AR), an ecosystem restoration initiative dedicated to the reanimation of Australia's arid lands, is about 400 kilometres north-west of Boolcoomatta and 130 kilometres east of Bon Bon Station Reserve. Established in 1997, AR has fenced 123 square kilometres of property, and feral cats, rabbits and foxes have been eradicated from a smaller area of 60 square kilometres, enabling AR to reintroduce four species of locally extinct mammals. All have reproduced in this secure environment.

At a Christmas party, Norris and AR staff discovered they each had similar ambitions to expand beyond their boundaries. Norris, at the time still reserve manager at Bon Bon Station Reserve, suggested a joint partnership that might include the local organic beef business and BHP Billiton, whose work in the region supports good conservation outcomes. He visualised others joining the initiative to create a loose group of partners to link up with Bon Bon and thus create another anchor region for Bush Heritage. O'Neill was enthusiastic, Bush Heritage signed an MOU with AR, and Norris and AR have jointly employed ecologist Aaron Fenner, a specialist in arid landscapes.

TOP Ethabuka Reserve, Qld, 2003

Photo: Murray Haseler

BOTTOM Taken in exactly the same location as the photo above, this image demonstrates the transformative effect of a new land management regime: Ethabuka Reserve, Qld, 2011.

Photo: Murray Haseler

In 2013–14 the project took shape, ticking all the boxes in the BioPrEP conceptual framework. These included a regional 'engagement strategy', 'increasing Bush Heritage land under management', 'encouraging more animals back on to Bush Heritage land', and 'more research projects with universities in South Australia'.[34] Norris's energy is infectious, and even the neighbours, conservative farmers and traditionalists, are interested, because he discusses fox baiting and weed control in ways compatible with their own aspirations. Slowly the community is pulling together to achieve conservation of the fragile ecosystems, necessary for both farmers grazing sheep and conservation managers.

Luke and Fiona creatively combined their several interests by bringing school groups out to the property and inviting Traditional Owners to tell their stories to the children.

Back at Boolcoomatta, the neighbouring farmers have noticed increased vegetation in certain areas of the reserve. As farmers do, they watched Ashton mapping 6000 rabbit warrens, and Norris's actions as he followed through with an eradication program. And they are moving towards cooperative scientific monitoring of feral pests, hoping that their pooled efforts will produce mutually beneficial results – a change from the first hostile meeting with neighbours when Bush Heritage bought the property in 2006.

Norris continued the work Ashton began, eliminating the artificial water points, and new plant and grass growth is occurring on the plains along paths of natural water flow after rain. He also rounded up the remaining rogue goats on Boolcoomatta, in the process discovering several small mulga trees pushing up between rocks. Not many, and not yet safe from rabbits, but they are the first sign that the trees that once covered the rolling hills are returning. Now as healthy landscape manager for the region, Norris's responsibility includes oversight of both Boolcoomatta and Bon Bon, and the work of the respective reserve managers. He and Luke Bayley, as healthy landscape managers, drew inspiration and enthusiasm from sharing stories and plans.

Bayley arrived at Bush Heritage by a circuitous route. As an eleven-year-old who enjoyed camping trips, surfing, fishing and bushwalking, he became interested in the environment when he learnt about the detrimental effects that chlorofluorocarbons were having on the ozone layer. He wrote to Joan Kirner, then Victorian Minister for Conservation, Forests and Lands, expressing his concern. Her reply encouraged him to continue to explore environmental issues. He went on to complete two degrees in Natural Resource Management.

In his search as an adult for a better quality of life with his family, Bayley believes that the natural environment is a crucial element. He and his family lived in Port Keats in the Northern Territory, where he worked with local Aboriginal people to build a ranger program. He was 'blown away' by the work, meeting elders, exploring magnificent remote country, and knew then that he 'had this outback thing in my soul'. They moved to Geraldton in Western Australia where he headed up the Ecologically Sustainable Rangeland Management Program for the state Department of Agriculture, work that involved a mix of ecology, land use planning and government incentive schemes for pastoralists. However, circumstances led to a return to Melbourne. But he disliked wearing a suit, and working in policy, so went back to the bush, managing the restoration of Lake Mokoan between Benalla and Wangaratta in Victoria for Winton Wetlands, and working on water rationalisation programs during drought; a great job, but he was still not entirely happy. When the position as reserve manager on Charles Darwin Reserve was advertised, he and Fiona decided to 'give it a crack', because they were looking for a 'more proactive, creative and autonomous' life together'.[35]

But upon seeing the property, Bayley was unsure. The reserve was dry, and semi-agricultural, not the rangelands he had envisaged. Was it really a good move to leave a $20 million project to settle in the backblocks of Mid West Western Australia? It turned out to be 'the making of us', and he feels 'blessed' to be there. Bayley inhabits the land as if it is another skin. He and his family entrenched themselves in the local community in Perenjori, the nearest town, 70 kilometres away, Fiona as a school counsellor and their children, Tanami and Banjo, at the primary school. Fiona loved the place and her work, and the children relished the freedom to explore. It was as though they had never lived anywhere else. Luke and Fiona creatively combined their several interests by bringing school groups out to the property and inviting Traditional Owners to tell their stories to the children. Ashley Bell, a Traditional Owner, enjoys these expeditions where he demonstrates how, long ago, his ancestors would widen and deepen the holes in the rocks to collect and store water by heating the edges around the hole with burning sticks until they cracked. He shows them how to treat the water with respect, since it is a scarce commodity, and helps the young Aboriginal children connect with their culture.

In 2015 Bayley took over Dave Whitelaw's position as regional manager of the West. With regret the family will move to make their home in a more central location since the job will require him to travel from Kalumburu in the northern Kimberley region to the Gondwana Link project in the south-western region. The Bayleys leave Charles Darwin Reserve knowing that it has staunch friends in Charlie Nicholson and Len Warren, who watch over its progress. Nicholson's comprehensive study of the property has been enhanced by Warren's more recent overview of the reserve's woodlands that he completed for Bush Heritage as a volunteer in 2012.

Warren's introduction to Bush Heritage began when his brother donated to the organisation in his name as a birthday present. The gift piqued Warren's interest, and he and his wife, Valerie, began donating, signing up for a donor's visit in 2003 just after Bush Heritage took possession of Charles Darwin Reserve. They were impressed by staff enthusiasm, and discovered kindred spirits among the volunteer group. When he retired in 2004 from his CSIRO job as a scientist in the mining industry, Warren embraced his environmental side, undertaking the review project at the reserve. He contacted botanists, discovered the Western Australian Herbarium, visited the Department of Environment and Conservation offices and the forestry department to collect useful information. He and Valerie spent weeks mapping and measuring each woodland, the diameter of each tree canopy, even locating tree hollows, fallen branches and leaf litter. He noted the relative intactness of the algal mats, the earthen crust made up of cyanobacteria that, together with tiny mosses and lichens, protect the ground surface from erosion and the drying wind. Using data collected over several years, he assessed the health of each woodland on the property.

The report highlights why protecting and maintaining woodlands is essential. As a macro ecological environment, a healthy woodland contributes to agricultural productivity by providing habitat for birds that pollinate plants, and for small marsupials that control insect pests. A healthy woodland protects against soil salinity. Destruction of woodlands has continued over 200 years of white settlement, leading to the loss of species and proliferation of invasive weeds. One-third of the ecological communities officially recognised as 'threatened' under the 1999 Australian government's Environment Protection and Biodiversity Act are in woodlands.

The Warrens host monthly meetings in Perth for Charles Darwin Reserve supporters, at which plans for week-long weeding parties are hatched. The first managers did not spray, and after ten years of dedicated weed-pulling by a group of 'amazing' volunteers, the property is regenerating, native grasses are re-establishing, and instead of fields of weeds, it is the individual weed that catches the eye.

At Scottsdale

I am on the landing at the back of the Scottsdale woolshed, breathing in Monaro Plateau air, clear and fresh, and feeling a touch of nostalgia. It is ages since I lived in Canberra, and I have forgotten how early sunlight pales the tussocky plains on still autumn mornings. On the embankment beyond, the historic railway line that once linked Goulburn to Bombala disappears into an elegant, oval-shaped, brick opening.

The tunnel, beautiful and undamaged after a hundred years of wear, cuts through the hillside down which a stand of eucalypts, small round leaves gleaming gold and silver, straggles untidily towards lower ground. Cliff Grant and I watch through binoculars a brown falcon that has alighted on a pole close by. Cliff cannot contain his satisfaction. A volunteer from Brisbane, he spends two separate weeks each year at Scottsdale, spraying weeds, planting, doing whatever is needed. He is a keen birdwatcher, and has nagged manager Peter Saunders for several years to erect a telegraph pole in the yard below the shed. Peter has recently done this, but until now the desired objective had not been achieved. Through the binoculars we see the white speckles on the falcon's breast and the steely, unblinking eyes surveying the surrounds. 'It's about rabbits,' Cliff remarks casually. 'If you've got a perch high enough, the falcon will watch for rabbits and they'll stay away.' Peter joins us, pleased that a falcon has settled on the pole while Cliff is here.

The easy banter between the two men shows how things work at Scottsdale. Peter is a man of silences and short sentences, an introvert at heart, although life at Scottsdale has taught him other ways to be. As we drive around the property, I understand that Scottsdale has been the catalyst for skills that Peter never thought he had; skills in building relationships and inspiring people to believe that restoration and revegetation can be achieved at Scottsdale, where 300 of the 1328 hectares were grazed, cleared, sown and cropped over four generations, leaving a battered and scarred landscape.

That evening I take to the river with Cliff, Peter and Brett Peden, Scottsdale's field officer. We putt putt up the river to the first long sandbank and sit on the warm sand in the gathering dusk with our beers, as craggy cliffs scattered with tough trees and grasses bow over the mirror-still river, a silver gelatin negative, the reflection almost more incisive than the real thing. Two black wallabies, disturbed, bound up the vertical cliff dislodging only one stone that tumbles with melodic clattering into the water, spreading a fan of ripples.

Innovation at Scottsdale

Scottsdale was purchased in 2006 as the anchor property that would ignite interest from local communities in the much bigger Kosciusko to Coast (K2C) conservation project. Although still an important part of K2C, in just a few years Scottsdale has become one of the flagship properties for Bush Heritage.

THIS IS PARTLY DUE to its location, 79 kilometres south of Canberra on the Monaro Highway. It is easily accessible for supporters and interested visitors who can take a day trip to participate in conservation activities. It is also due in part to the ambience created by its reserve manager, Peter Saunders, who practises conservation in an innovative and imaginative way, drawing those around him into his grand plans for the property.

Saunders is a former chartered helicopter engineer specialising in maritime aviation, who then turned to farming and obtained an environmental management degree. In 2009 he became manager of a landscape property that is home to some of Australia's most threatened temperate ecosystems, in particular the mallee silver-leafed gum, that survives today in only ten small populations, and is listed as vulnerable to extinction.

At the time of purchase, Scottsdale was degraded from overgrazing; paddocks were full of invasive weeds, mainly serrated tussock and African love grass; erosion in the gullies was cutting back into the creeks, and spreading. Bare hills should have been covered with endemic woodlands, and invasive poplars were increasing on the valley floor, forcing out all but a few small patches of native grasses; willows choked the water flow in the Murrumbidgee River. The aim was to restore an entire ecosystem, by encouraging natural regeneration and replanting key species found in yellow box woodland. Work on Scottsdale that demonstrated effective restoration was essential for the success of K2C. Although some work had already been done, Saunders, when he first surveyed the property, remembers thinking, 'What have I done? I can't do this, I'm only one person!' This was soon followed by the realisation that it was people, over 150 years of them, who had degraded the land, and that it was up to people to restore it.[36]

He began working with the Ngunawal people on a comprehensive cultural heritage survey to understand which sites needed special protection. They unearthed artefacts, stone axes and tool-making materials, and through this joint experience, Saunders and the Traditional Owners gained confidence that an ongoing dialogue would benefit the long-term restoration of the property.

Being an engineer, Saunders looked to the cause of the problem for an opportunity. The solution was in Canberra, only forty-five minutes drive away. Canberra had an unusually high number of Bush Heritage donors; people power could help him achieve the plans he was formulating. But first he had to create an amenity for volunteer groups so that they could enjoy the experience. The historic galvanised iron woolshed, a charming feature of the property, had been the location for the film *The Sundowners* in 1961, but now it was dilapidated and had no services. In 2009 Bush Heritage carried out a full refurbishment, creating a kitchen in one corner of the shed, with drinking water, and toilets were installed close by. The next step was to work with the fundraising–marketing people from Melbourne on material to send to Canberra supporters. Would they read it? Would they be interested? Would they come?

A single *Callitris* (native conifer) stands on a slope, its unusual rounded canopy putting its age at 150 years. It is a lone survivor from the time when the hillside was a forest of *Callitris*.

They came, they participated, they told their friends, and returned again and again. In the 2014–15 financial year, volunteer hours reached 4200, two and a half full-time employees worth, and much has been achieved. Managing volunteer expectations and matching people to tasks is unexpectedly enjoyable, Saunders says. Words flow when he talks about the property, and volunteers are inspired by his experimental and collaborative approach to conservation that is starting to show exciting results. Local farmers are visiting, too, taking back to their own farms ideas they have seen working at Scottsdale; and Landcare groups and others interested in bushland connectivity are learning the skills that will enable them to join the K2C project in their own areas.

In autumn the African love grass sports fluffy blonde heads, and seems to be everywhere. But several paddocks are already partially cleared. Native grasses are predominating again, enough to reintroduce the striped legless lizard, which is thriving. Restoration and continual management of grasslands is difficult and expensive, however. The quest, Saunders affirms, is for 'balance' in the landscape: he is working towards regenerating sufficient red-leg and kangaroo grass (natives) that will hold their own against the love grass, which is impossible to annihilate. Shade inhibits its growth, however, and Saunders is controlling the weed with systematic, well-directed spraying, and judicious planting of trees. In a fenced-off patch of bare ground, Saunders has, experimentally, taken off the top 10 centimetres of nutrient-rich topsoil in which the current love grass seed rests. Into the bare ground he has sown twenty varieties of native grasses, thousands of seeds, and he watches their progress with the prideful eye of a parent. Small clumps of green are beginning to surface, and Saunders hopes that they will flourish and that, helped by the wind, their seed will spread further afield, a natural agent in the quest for balance. Saunders also welcomes the scientific and experimental research being conducted by a PhD student in several fenced-off plots on Scottsdale. The student is using satellite imagery alongside ordinary photography to measure the growth of different grasses in different situations on a number of properties, and results will help direct further improvements that Saunders has in mind.

The freedom to experiment, coupled with an engineering background, has paid dividends, as has his collaboration with Paul Gibson Roy, from Greening Australia. Together, they have built a machine that 'just tickles' the soil, and plants seeds of native grasses and trees in strips of wet sand. Volunteer Suzanne Connelly leads willing helpers to collect native grass seeds and grassland flower seeds, a task that would otherwise be prohibitively expensive, so that Scottsdale now provides

Reserve manager Peter Saunders and researcher Brett Howland work on the reintroduction of the striped legless lizard. A dynamic volunteer program is now transforming the Scottsdale landscape.

Source: Annette Ruzicka

most of the material for its own regeneration. Connelly, who lives 6 kilometres away in Bredbo, manages the Scottsdale plant nursery, even coming in on Christmas Day to water her plants. In her sixties and a lover of native bushland, she has taught herself the botanical names of all her charges, and learnt the skilful arts of propagation and seed collection. Dedicated and enthusiastic, her physical and mental wellbeing are now inextricably bound to that of Scottsdale.

On the hill, amid cream scribbly gums and other trees comprising the grassy yellow box woodland, Saunders can look down on the scarring of the valley floor, where direct seeding has followed the contours of the bare ground in three long lines, 6 metres apart to allow for spraying. The trees are small but flourishing, and in time, Saunders hopes, the furrows will disappear. Dotted about the hills are small copses of trees, hand-planted, to give 'structure' to denuded slopes, without obscuring the salient native grasslands. In three years, over 100 000 tube-stock trees have been planted, testament to a powerful volunteer program working alongside the direct-seeding project and volunteer weedsprayers.

A single *Callitris* (native conifer) stands on a slope, its unusual rounded canopy putting its age at about 150 years. It is a lone survivor from the time when the hillside was a forest of *Callitris*. Saunders checks the small bush-like trees around it that are *Callitris*, too. Post-grazing, the species is returning, another significant step.

From a high lookout point, the Murrumbidgee is a silver ribbon trailing through the landscape. On the way to the river, the track crosses gullies where engineered earthworks, not yet covered in vegetation, reveal recent efforts to halt erosion. In the main creek, a handsome steel and rock wall makes an expensive and effective attack on the erosion that threatens the health of the property. Rock pools stepping up the slope create a 'fish ladder' that enables free movement for fish, but halts the unravelling of the soil in times of flood.

Antia Brademann is the facilitator of the Upper Murrumbidgee Demonstration Reach (UMDR), the goal of which is to recover native fish populations and improve river health in the Upper Murrumbidgee. It is an ungainly title for an exciting project with real energy, a dynamic leader, and many supporters. Originally funded by the

Murray–Darling Basin Authority as part of its Native Fish Strategy, the river reaches have been used to investigate the different interventions that degrade the habitat and quality of water, information that can be used to help in fish recovery. When the Reaches program lost funding, it was reinvigorated by the Scottsdale initiative as part of the K2C project. Because Scottsdale has a long Murrumbidgee frontage, making it part of the catchment area, Bush Heritage stepped in to partner UMDR, providing funding, greater credibility, and accommodating a proposal by PhD student Prue McGuffie to study the habits of the endangered native Macquarie perch. Thus, the interests of several different groups have combined to develop a much bigger project that has now called for comprehensive monitoring of the river's fish inhabitants. The aim is to understand all fish movements and the role that the 'evil' interloper, the European carp, plays in suppressing native fish numbers. It is believed that while carp are not directly responsible for the decline of native fish, they are a likely factor in keeping numbers low, and the project studies the conditions where this might be true. Saunders has acquired a carp-trapping licence, necessary for compiling verifiable numbers and results. Fish of all kinds have been caught and tagged, and monitoring equipment in the water has captured fish movements for analysis. The project continues with donor and volunteer support, and the new data and interim findings from McGuffie's research, some of which contradicts current knowledge on the subject, will have relevance for other inland rivers.

Saunders works closely with ecologists, combining regeneration and restoration work with his own creative and innovative practices learnt earlier in his engineering life. He thinks nothing of developing a piece of machinery to help him achieve a better outcome. But it is the enjoyment he gains from working with volunteers and supporters that sustains him. Because he has experienced it himself, he can see how those who come to Scottsdale arrive at some measure of peace and happiness after a day's hard work in the natural environment.

Scottsdale Reserve, NSW, is undergoing a transformation through the work of volunteers.

Source: Jiri Lochman/Lochman Transparencies

The highs and low of 2012

At the beginning of 2012, the Gondwana Link project received endorsement from a Queensland couple, Bill and Jane Thompson, who sold their 10 hectare custard apple farm in the Glasshouse Mountains in south-east Queensland to buy the 1500 hectare property known as Yarraweyah Falls that adjoins Monjebup Reserve in the south-west of Western Australia.

THIS LINKED MONJEBUP to Monjebup North, completing a U-shaped area of 3000 hectares of continuous native habitat.

The Thompsons had long been interested in conservation, having restored their own small property to nature refuge status, and when the Queensland government compulsorily acquired their land for railway infrastructure, they put their interest in conservation to a bigger test. On a trip to Western Australia's south-west region, they met Keith Bradby and Simon Smale; inspired by their ambition and vision, they decided to buy into Gondwana Link, with active management support from Bush Heritage.[37] The Thompsons have settled into their new environment, with its different climatic conditions and vegetation, and Jane has amassed a 'pretty detailed herbarium' while collecting information on the local plants. In 2013, after harvesting seeds themselves from their property, they direct-seeded 100 hectares of land in conjunction with Bush Heritage. Vegetation surveys undertaken with Bush Heritage have already yielded 600 species, but they have not been as lucky with fauna, since the wildlife still comprises mostly foxes, cats and kangaroos, although 'night critters' do include six species of bats. The Thompsons hope that, in time, more birds and small mammals will return to Yarraweyah Falls, and that their work in promoting the health of the region will inspire others.

As already mentioned, after the GFC and up to the start of 2012, Bush Heritage had retreated into itself. These were years of 'taking stock', coming after the hectic years of land acquisition. Systems underpinning the enterprise were finetuned, in particular the financial basis on which the organisation rested; the Miradi system, connecting all aspects of the conservation business, and the Health and Safety and Environment Management system were in place, and working well with general approval;

partnerships, including those with Aboriginal groups, university research groups, conservation groups, and local community organisations were developing in many of the priority regions, and communication between the Melbourne office and the outlying reserves was strengthened. BioBlitzes had been conducted on most of the properties to ascertain the base ecological health of the landscapes, and results were analysed by the ecologists, university departments and science groups involved, and fed into the central system, enabling more detailed and precise planning on the macro as well as the micro level. And Gerard O'Neill's appointment towards the end of 2011 had sparked new energy. Once again, Bush Heritage's thoughts were tending towards land acquisition.

Arriving in 2011, O'Neill was conscious of Bush Heritage's 2025 goal to protect 1 per cent of Australia's land surface, and he was also keen to make his mark with a property purchase. He therefore welcomed the request from the Environment Water Trust (EWT) and Professor Richard Kingsford, director of the Centre for Ecosystem Science at the University of New South Wales, that Bush Heritage consider buying Naree Station from Paul and Debbie Kaluder. Tucked into the remote north-western corner of New South Wales, Naree Station is a pastoral property in one of the least disturbed parts of the Murray–Darling Basin. It comprises 14 400 hectares of flattish land sheltering creeks, claypans and wetlands, deep in mulga country, 150 kilometres north-west of Bourke, on the Cuttaburra Channels connecting the Paroo and Warrego rivers. Its southern boundary is the Cuttaburra Creek, and Yantabulla Swamp adjoins it on the west. Subject to the 'boom and bust' cycles of wet and dry that are typical of the rangelands of inland Australia, the erratic rainfall, averaging an annual 300 millimetres, means that the property is dry for long periods. When it rains, Naree transforms into an exquisite and intricate wetland of diverse water environments, each attracting different types of birds. The Kaluders' six-year care of the fragile environment, and their innovative and progressive grazing methods, had partly restored the heavily degraded property. They, with Kingsford and the EWT, wanted their unique property to be protected and nurtured forever. Jim Radford, now Bush Heritage's science and research manager agreed, describing it as 'part of one of the twenty most important wetlands in Australia'.[38] Bush Heritage ownership seemed the best option for this outback wonderland. At the time Bush Heritage was contacted, there was a big risk that the property might be bought by a pastoralist whose sole focus would be agricultural production. Bush Heritage had to act quickly and bought the property before presenting the project to supporters. They also secured David and Sue Akers, experienced conservationists, as Naree's first reserve managers. The Akers had spent years in the Channel Country, that remote corner where Queensland, New South Wales, South Australia and the Northern Territory meet, but this region was home to David, who had originally farmed a family property here.

Donors invited to discover Naree Station were overwhelmed by its magic, and generously supported the project. At the time, Naree Station was presenting in all its glory. Following Cyclone Tasha in 2010, and Cyclone Yasi in February 2011, both of which devastated the Queensland coast, Naree had burst into life. Local rain and run-off coursing down the Warrego River filled every waterway. Over 170 species of birds descended on lakes and wetlands teeming with fish and other water creatures, or to gorge themselves on the nectar and fruits from flowering trees and plants. The relatively mild temperatures at Naree make it an important migratory destination for birds from colder climates, and with water everywhere they came in their thousands to feed and to nest in the trees that fringed the water-filled creeks, semi-permanent waterholes and ephemeral lakes. The property glowed with bursting, fecund health, alive with birdsong and flight, the swish and plop of fish, and snakes and other reptiles, and at least fifteen mammal species out

FOLLOWING SPREAD **Naree Station wetlands are a magnet for waterbirds in wet years.**

Source: David and Sue Akers

and about, all chasing food and mates, and endeavouring not to end up as another creature's meal. The brolgas, in elegant rituals designed to attract a partner for life, caught the eye of Stacey Irving, newly employed at Bush Heritage: 'With wings spread wide, the silver brolgas jump, dance, pirouette and prance on their long, stilt-like legs, [all the] while making loud trumpeting calls to each other. They beat their wings, and bow and bob their red-banded heads – all in the name of love'.[39]

Conditions are not always so rosy, however. The rain stopped, the birds flew to wetter places, the frogs retreated into the drying mud, the trees and bushes drooped, bracing against an unknown timeframe of drought. Undeterred, the Akers implemented conservation plans, fixed fences, trapped and removed feral goats and pigs, and settled into the community, making friends and common cause with their neighbours, preparing for the next 'wet'. They also talked with the Traditional Owners, the Budjiti people, who have maintained close personal connections to the property. Phil Eulo and his family began working with the Akers, bringing the history of the place alive for them as they toured the property together, assessing its environmental and heritage values. Their efforts highlight the long-term, collaborative approach at the heart of Bush Heritage's conservation strategy. For Eulo it is a big step forward, too: 'Bush Heritage is doing what we wanted to do all along – keep our country natural. Now we've got the opportunity to bring this back to its natural state … for new generations, white and black'.[40]

On the other side of the country, in Western Australia's Kimberley region, on a bright day in the early dry season in October of 2012, healthy country manager Tom Vigilante undertook his first 'fire walk' with the Uunguu rangers in Wunambal Gaambera traditional lands, part of the healing and refreshing of *uunguu* – their living home' – that must have taken place each year for centuries, to protect the area's vulnerable animals, plants and vegetation, and prevent destructive wildfire, through 'right-way burning'. The walk took five days and covered 50 kilometres in a remote and rugged area of the Kimberley, through sandstone gorges that bite deep into the coastline, from the tops of cliffs down to inland flat plains, and along the Hunter River, ending on the Mitchell Plateau. They walked as a team, lighting fires in the traditional way as they followed the creek line, watching for wildlife and recording all sightings of animals, plants and birds on hand-held data-collecting GPS devices, their daytime activities a happy marriage of ancient technique and modern technology. They also recorded the route they took, and the cultural sites, noting the presence of feral animals and cattle, data to influence their plans to build fences to protect native wildlife and rock paintings. They hunted freshwater crocodile and fished for bream, and collected bush foods and medicinal plants. The walk was another step forward in the sharing of two-way knowledge that is continuing to develop between the Wunambal Gaambera people and Bush Heritage.

If 2012 saw several milestones achieved, it ended on a terrible low with the death of a much-loved employee, 25-year-old Mauritz ('Mo') Pieterse, while working with Bush Heritage. Pieterse was living at Cravens Peak Reserve, working as a field officer and offsider to reserve manager Peter Welldon. Originally from South Africa, he had found his ideal job with Bush Heritage, working in conservation on the edge of the Simpson Desert.[41] The effect on the close-knit community that is Bush Heritage is hard to overestimate. A fun-loving person with a gift for making friends, Pieterse was well known and well loved in the short time he worked for Bush Heritage, and respected for the knowledge and understanding he had of his environment. Colleagues were left deeply saddened by his death.

Gerard O'Neill, as Chief Executive, was devastated by the event. A tragedy of this magnitude, coming so early in his time as leader of the organisation, was especially difficult. The events at Cravens Peak would cast a shadow for some time, with closure for many coming only a year later with the creation of a fitting memorial to a good friend and colleague. It was erected at Ethabuka at the place where Pieterse had died, and a moving ceremony was held to commemorate his life.

Science underpins a bold future

On 13 June 2015, Phillip Toyne died after a long battle with cancer – a huge loss of a powerful advocate for the conservation movement. For Bush Heritage it was a time to celebrate the achievements of the man who had done so much to lead the organisation through great change into a bold future.

AND AT THE END OF 2015, as Bush Heritage approaches its twenty-fifth year, the changes in direction that he largely instigated are becoming clearer. As government funding for land purchases dries up, it has made sense to ensure Bush Heritage's ongoing financial stability by restructuring the Endowment Fund, now the Capital Fund, to secure management costs in the long term and to support land purchases and the more intangible issues that come with partnership work. Today, the focus is on reserve and partnership networks, and priority landscapes, and efforts are directed towards consolidation in these areas. Confidence in these new directions has grown, and the concepts that were born as part of the fledgling Beyond the Boundaries concept are now mainstream for Bush Heritage and its supporters.

Sue McIntyre, senior principal research scientist with CSIRO Sustainable Ecosystems, observed the changes after joining the Board in 2007, in the midst of the frenetic land-buying period. Her own background fits her well to participate in the philosophical discussions that have occupied the Board more recently. After growing up on her family's wild 4 acres of land on the Yarra River in Kew, in Melbourne, her fascination with nature led to a deeper interest in botany in the mid-1980s, when ecological studies were linked with agricultural production. Before joining CSIRO, she worked for Land and Water Australia, brokering research relationships between industry and academia by communicating scientific research and by physically working with the public, farmers and landholders to achieve results. A specialist in weeds and grasslands, she stepped into Louise Gilfedder's role on the Board and with Steve Morton, another CSIRO scientist and her boss, helped concentrate the Board's collective mind on issues of effective management, particularly of grasslands and woodlands, which are now seen as some of Australia's most threatened landscapes. With Chief Executive Gerard O'Neill and President Louise Sylvan spearheading new approaches, Board discussions have centred around just how much intervention is necessary to maintain a property, and what science is available to ensure success. Discussions with Bush Heritage ecologists have come to the bleak but liberating conclusion that there are no definitive answers, and that each property is unique in its requirements. This has provoked new questions as to what can be done to facilitate the processes on which successful conservation depends, and the answers all come back to the need for more science, and for greater knowledge and understanding of the ecological processes that underpin conservation efforts. The Board has also considered how it can support a greater science capability in applied ecological research, and what research themes would yield the best results. Working with the Bush Heritage Ecology team, they have decided on six themes: landscape connectivity, threatened species management and recovery, habitat refugia, fire ecology, restoration ecology, and introduced species and over-abundant native species. From this, an ambitious project that has become Bush Heritage's ten-year Science Plan has been born. It was launched at Scottsdale in March 2015.

In announcing its Science Plan, Bush Heritage drew attention to the extinction crisis for thousands of species. As one of seventeen mega-diverse countries in the world, with 7 per cent of the world's remaining species, Australia's obligation to ensure their survival is pressing.[42] The Science Plan presents a powerful case for ramping up scientific efforts to protect the numbers of plant, land mammal, bird, reptile and amphibian species. When Bush Heritage announced financial support for the Science Plan, it reaffirmed its commitment to, and dependence on, science for making the right decisions using existing knowledge, for generating new knowledge through research and modelling, and for improving the design of biological surveys and ecological monitoring.

The ambitious Science Plan includes fifty-five active conservation research projects, with designated science hubs on key reserves, involving fifty scientists from fifteen universities. The plan also includes the potential to double Bush Heritage's science capacity by 2025 through collaborative research, science fellowships, and citizen science. There is support for young scientists, too, with new postgraduate scholarships, internships and placements, and enough growth to support 120 collaborative research projects aimed at halting biodiversity loss by 2025. These initiatives will bring a compelling impetus to science-based ecological research throughout Australia.

The plan builds on much of the work already in place on the reserves, and on connections already cemented with academic research groups. And, in bringing it all together, Bush Heritage makes a strong case for positive connections with all agencies involved in conservation, and for a concerted effort to pool knowledge and expertise. The launch at Scottsdale was well attended and received multiple media coverage across the continent and, pleasingly, a subsequent increase in first-time donors.

The Science Plan is one of several initiatives that came to fruition in 2015, and it is possible that Bush Heritage has never been busier. At much the same time that Scottsdale was humming with visitors and guests, other Bush Heritage staff were progressing a new partnership arrangement with the Arafura Swamp Rangers Aboriginal Corporation (ASRAC) in Central Arnhem Land. They visited Ramingining to hold pre-planning and governance meetings after ASRAC invited Bush Heritage to be a key partner in supporting its management capacity. Unlike some other groups, the people of the Arafura Swamp have maintained strong continuous connections with their land. Many still live on country, which has facilitated a strong consensus approach to developing a Healthy Country Plan. When this has been combined with visits from Bush Heritage personnel from Melbourne who have particular expertise in governance and administration to help build ASRAC's capacity, and that of the community-based rangers, there have been positive results. Emma Ignjic feels that Bush Heritage is at last pulling together a solid team to create viable partnerships. This agreement with ASRAC, when combined with the Warddeken IPA, the neighbouring Djelk IPA, and Kakadu National Park to the west, encompasses a substantial protected area of more than 4 million hectares of Aboriginal land in the Top End, where Bush Heritage has a unique opportunity to support and influence conservation outcomes in the region.

Other partnerships have progressed, too. In December 2014 the Olkola Aboriginal Corporation (OAC) on Cape York celebrated the return of five former cattle properties, which now makes them one of the largest non-government landholders on the Cape York Peninsula, owning a total of 869 822 hectares. The Olkola have progressively been coming home to live on their land since Glen Garland Station was returned to them in 1996, and the Alwal National Park in 2010. During this time they have practised land management in the traditional ways of their forefathers, ways that were in place well before white settlement. Their aim has been both to conserve the land and to develop economic possibilities so that more Olkola people can 'come home'. One of their first activities when they began working on a Healthy Country Plan in 2012, was to create training programs for land managers and in 2015 they are justifiably proud of their highly skilled band of enthusiastic land managers. They have developed extensive weed control projects and their traditional fire management work is one aspect of their eligibility for the Carbon Abatement Scheme. As OAC's programs have developed, they have pursued partnerships on an equal footing with organisations they believe will support their goals of protecting Olkola cultural heritage and building economic opportunities on the enormous tract of land that requires appropriate income sources. In 2014 the Corporation invited Bush Heritage to support Olkola's strategic development and conservation activities. An MOU was drawn up and signed towards the end of 2014, and in 2015 Bush Heritage was able to play a role in funding the finalisation of the Olkola Healthy Country Plan and in providing expert personnel for particular projects, to support the successful work that has been ongoing for a number of years. Olkola land includes an important range of ecological systems of high integrity and good connectivity, and seasonal refugia that are associated with the wetlands, riparian environments, and critical habitat, and Bush Heritage's years of work

Olkola land managers and Bush Heritage ecologist Allana Brown are sharing two-way knowledge as they work on saving Alwal, the endangered golden-shouldered parrot.

Source: Allana Brown

protecting and conserving similar landscapes and its contribution of scientific expertise, have augmented the capacity and experience that Olkola already have in directing and managing their country. Partnering Olkola in the conservation management of approximately 800 000 hectares of land also helps Bush Heritage achieve its vision of influencing the protection of 1 per cent of Australia's land surface by 2025.

Olkola's flagship project is to secure the future of the golden-shouldered parrot. They believe that Bush Heritage's commitment to landscape-scale habitat management will complement their own skills, and that by teaming up with two parrot experts and neighbouring landholders, they will have a good chance of saving the bird. The golden-shouldered parrot, a jewel-bright, budgerigar-sized bird, lives on seeds in the grassy woodlands of the peninsula and nests in anthills in the savannah land. Once prevalent throughout the southern Cape York Peninsula, it is recognised as endangered by both the Queensland and Australian governments, being now restricted to two small areas covering less than 2000 square kilometres. The parrot, or *Alwal* in Olkola language, is of great cultural significance, being a totemic species for the Olkola, and as they now own about 60 per cent of the species' remaining habitat, they are putting maximum effort into ensuring its survival.

Partnering Olkola in the conservation management of 800 000 hectares also helps Bush Heritage achieve its vision of influencing the protection of 1 per cent of Australia's land surface by 2025.

The key threats to *Alwal* were identified and ranked during a team planning workshop with Olkola land managers and Bush Heritage staff, and the degree to which each threat impacts on the parrot and its habitat guided the planning and prioritising of the action plan that will be implemented over a five-year period. Strategies for the bird's protection and increase in population include restoring appropriate fire regimes to promote preferred grass food species that should improve breeding success, controlling pest fauna species to improve *Alwal* habitat conditions, addressing the critical food shortage period in the late wet season, and monitoring progress over five years. By this time it is hoped the bird will be prevalent in such numbers as to delight both Olkola people and those lucky enough to experience ecotourism on Olkola land. *Alwal* is an important rallying point for the Olkola and one aspect of the project is devoted to capturing the stories and traditional knowledge about *Alwal* that will build understanding for visitors to the region of *Alwal*'s significance to the land and its people. For Bush Heritage, this is an important project in which the health of the landscape will determine the outcome for this beautiful bird that now has every chance of being saved.

A poignant moment that captures the energy of the Bush Heritage/Olkola partnership occurred early in 2015. Glen Kulka, an Olkola land manager, has devoted himself to learning all he can about the habits of *Alwal* from his traditional knowledge and from fieldwork. To gain a better understanding of its daily habits, he closely examined some of the 20 000 photos captured by monitoring cameras of the birds arriving at and departing from the termite mounds. Using this modern technology with close analysis of the bird and its habits, he carefully noted the times they came and went, their activities, the gender and age of the birds, and developed the theory that the adult birds would likely be drinking at 2 pm. To test his hypothesis, he and Bush Heritage ecologist Allana Brown positioned themselves at the nearest waterhole and waited. A few discreet cheeps, soft small budgie-like noises, from the trees across the waterhole alerted them to the arrival of an adult pair of *Alwal*. There was a flurry of turquoise, iridescent green and flashes of gold as the birds descended, their black and white heads dipping as they took a quick sip from the water before flying off. Brown hit the clock on the dashboard – it was 2 pm exactly! Even though the birds were only in view for about two minutes, it was enough to prove the point, and Kulka and Brown were ecstatic. 'We were so excited, we couldn't stop giggling like school kids, and I had a grin on my face for the rest of the week, as it was the first time I had seen adult parrots close up for a decent length of time.'[43] Another step had been achieved in understanding *Alwal*'s movements that will surely help in its protection.

At Naree Station, in early May 2015, 30 millimetres of rain fell, not enough to fill the wetlands but enough to encourage bird activity, and to trace the trickles of water across the property as it puddles out, bringing a shallow covering to some of the ephemeral lakes, and new growth. Reserve managers Sue and David Akers explore their back country, and the intricacies of the waterways and ephemeral wetlands at such times, learning how they connect, an impossibility when the land is completely dry. They follow small channels that lead to real water, the massive Yantabulla Swamp. Unlike the Naree wetlands, the Yantabulla Swamp usually floods every three years from the overflow of the Paroo River and from other creeks that flow from the Warrego River. On the edge of the swamp, the Akers look across the marvellous expanse of water that stretches to the horizon. There is special pleasure, too, knowing that Bush Heritage is in the process of drawing up an agreement for management of the swamp, in partnership with the property's owners, South Endeavour Trust (SET).

The addition of Yantabulla Station to Bush Heritage's stable of management represents a conjunction of purposes for Bush Heritage and SET, a private fund that supports conservation objectives. With Bush Heritage, SET, headed up by Tim Hughes, had been considering projects that would benefit the landscape. When Yantabulla Station came on the market, they knew it was the perfect vehicle. SET could buy the property to protect it, and appoint Bush Heritage as manager, thereby creating huge efficiencies and cost savings, since Bush Heritage is already managing Naree Station next door. David and Sue Akers are excited by the prospect, and as Sue points out: 'Yantabulla Station has wonderful natural values that echo and complement what we have at Naree. The issues of feral animals, fire management and weeds are very similar to Naree, so we can hit the ground running with all the knowledge we have gleaned from our first two years here'.[44]

Bush Heritage's executive manager for the South-East Region, Jody Gunn, visited Naree soon after she started work in 2015. She sees the reserve as the building block for other partnerships to evolve, and in the transaction between SET and Bush Heritage, the possibilities for expansion of the Bush Heritage management model are apparent.[45] It is possible that similar extensions to Bush Heritage's land management responsibilities will occur in other locations, as adjacent land becomes available for sale, and Bush Heritage is seen to be a responsible and involved member of the local community.

On the other side of the country, on Charles Darwin Reserve in Western Australia, Luke Bayley, about to relocate, could look back on several years of productive involvement with local and more distant communities. Since the reserve straddles the divide between the arid west region and the more fertile south-west, it has always been a good place to study the effects of climate change, which everyone agrees are having an impact on the agricultural Mid West. The general trend for less rain in winter and more in summer has seen wheat production dwindle, and Perenjori Shire Council members support studies conducted on Charles Darwin Reserve in any way they can. Ecologist Nick Dunlop has maintained a climate change observatory there for ten years, and has seen the return of birds as the health of the land improves. Today, the splendid fairy-wrens that are captured are ones that have been captured in a previous trapping program, indicating that the birds are now resident. The trapping takes place on what was a hay paddock fifty years ago, and observations as to how the paddock has evolved in the past few years add to the data collected on the effects of climate change and how to combat it. Dunlop's program has involved hundreds of volunteers and undergraduate students from Environment and Zoology programs, keen for hands-on practical experience, and Bush Heritage's welcome has made this possible. Dunlop believes that long-term studies are essential to understand the effects of climate change on plants and animals, and that monitoring programs led by private conservation have the best chance of longevity.[46]

Volunteers, too, have provided practical work for Bayley's experimental project to grow more York gums on the property, and he has employed local farmers to assist. They have fenced off a 1 hectare block, ripped the ground and laid branches of York gum, complete with seed cases, across the block, judging that this method has the best chance for self-seeding and regeneration of this picturesque tree. In time, it is hoped that Charles Darwin Reserve will support more extensive York gum woodlands. Volunteers have also helped ecologist Vanessa Westcott in placing forty motion sensor cameras at strategic points to gather more information on the habits of feral cats on the property to help with eradication programs. Satellite technology, financed by the Gundawa Regional Conservation Association from its mining offsets, is adding to the knowledge of local communities about the elusive malleefowl, with the discovery of thirty nesting mounds in thick bush.

And in the early part of 2015, in Queensland, Steve Heggie, reviewed some of the projects under his care as healthy landscape manager, and decided to take drastic steps, first at Edgbaston and then at Goonderoo. The first attempt at a captive breeding program for the red-finned blue-eye was a disaster, as Adam Kerezsy had predicted.[47] The rationale for expecting these fish to breed in acquaria (deep water relative to the shallow springs) was flawed from the start. With Kerezsy no longer working for Bush Heritage, Heggie decided to create two artificial springs on the property to replicate the natural springs. He and volunteer Dave Coulton have created small mounds above the natural ground level and then trenched in water from a bore so it bubbles up through a riser, creating an environment as much like the natural spring as possible. Renee Rossini, the PhD student who works with the rare snail species, brought in fellow students to help seed the 'bubbles' with native aquatic plants and vertebrates. When the 'bubbles' are ready, there will be another attempt at captive breeding, with more success this time, Heggie hopes. As further insurance against damage to the natural

Bunuba partnership team. From left: (standing) Kendrick Chungal, Steve Heggie, Elaine Marr, Petrine McCrohan, Henry Williams; (seated) Nigel Hale, Natalie Dave

Source: Mel Sheppard

springs and the barriers protecting the fish colonies, he replaced 15 kilometres of fence 'pretty much totally reducing cattle from getting into the property'.[48]

Steve Heggie reviewed some of the projects under his care as healthy landscape manager, and decided to take drastic steps, first at Edgbaston and then at Goonderoo.

Goonderoo has always niggled at Heggie's conscience too, since the days when Bush Heritage reduced staffing to concentrate on the new and more demanding Carnarvon. The GFC gave Bush Heritage the opportunity to establish a cost-neutral program at Goonderoo, by leasing the house and agisting cattle to keep the buffel grass down, but Heggie always believed they could do more. He enlisted the voluntary help of Shane Jackson, a Queensland Parks and Wildlife pest control officer, and Chris Dickman from Sydney University, to carry out a comprehensive review of the pest management strategy at Goonderoo. At the same time, the pair used thermal imaging, trapping, shooting and sand traps across the entire bridled nailtail wallaby relocation zone to compile another report, and have since made several recommendations that will help Bush Heritage in its renewed quest to turn Goonderoo into a reserve of which they can be proud, after the years of relative neglect.

All this began a busy year for Heggie who, with partner Mel Sheppard, relocated to Fitzroy Crossing in Western Australia at the end of July. Having followed Bush Heritage's involvement in the establishment of successful partnerships with Aboriginal groups further north, the Bunuba people of the Fitzroy Crossing group asked Bush Heritage for help in developing a Healthy Country Plan for their country. Bush Heritage was cautious, suggesting that they would not consider the project unless the Bunuba had an operations manager to support governance, and that it would be necessary to embed someone in the community for two years to develop mutual respect in order to facilitate and deliver the plan. The Bunuba agreed to support both positions, and Heggie and Sheppard successfully applied for the roles of facilitator and operations manager, for a two-year period. A month into their new positions, they were getting to know their community, and Heggie was already arranging small meetings with different family groups to inform people about the Healthy Country Plan, in preparation for a big community meeting in October, so that a consensus could be reached on a community position that would direct the rest of the planning process. The hope is that there will be sufficient community cohesion and agreement on the directions to be taken before the wet season closes down much of the activity and operations for the year. For Heggie and Sheppard, the move and the work are challenging, and they have plunged in with the same enthusiasm that took them to Carnarvon years ago. In their spare time, they explore the magnificent country that is home to the Bunuba people, a land full of surprises, with high broken escarpments tumbling into deep gorges, rich grassy plains and pockets of forest on either sides of dry riverbeds that course with water pouring from the cliffs in the wet season. The Windjana Gorge, the Geikie Gorge and Tunnel Creek cut into the ancient barrier reef that is a rare remnant from the Devonian period, an imposing landmark redolent with the history of conflict when graziers were opening up the country. It was here that the Aboriginal leader Jandamarra led his people in battle for six years against the white settlers, using Tunnel Creek as his hideout, before being killed in 1897. For the Bunuba people, ecotourism is one of the opportunities they will consider as they work towards a Healthy Country Plan.

Flexibility in roles is a key attribute for Bush Heritage staff. In a swap that suits both parties because they are at different stages in their lives, Al Dermer has exchanged his role as healthy landscape manager in Tasmania to spend time as reserve manager at Boolcoomatta. It is a chance to 'go back to the bush' while his children are still young, while Glen Norris slips easily into the position of healthy landscape manager for Tasmania and Victoria. Both are demonstrating their desire to remain with Bush Heritage as long as possible.

Exciting times

While management on reserves and partnerships with Aboriginal groups have been progressing, Bush Heritage has not neglected its mission to purchase property. But the desire to own land just for the sake of the hectares it delivers has dissipated, with the wording in Bush Heritage literature shifting to convey the idea of protection and support.

THE BOARD IS NOT KEEN to buy property unless it meets stringent scientific guidelines, and has such outstanding potential that it would be impossible to ignore. In 2014 two such properties appeared on the horizon, and all efforts were put into acquiring them. Each property offers different opportunities and each, in its own way, holds a key to Bush Heritage's future that cannot be fully envisaged at this point. In a fitting summing up of its twenty-five years in conservation, these new purchases point more than ever to the long-term vision that has always been central to the organisation, because they will set a new agenda, testing and extending Bush Heritage's capacities well beyond the lives of those involved today. As this history closes, their acquisition, fittingly, leaves Bush Heritage's story in cliff-hanging mode.

The first of these properties is Hamelin Station in Western Australia. For those who have travelled to Shark Bay, the sign to Hamelin Station might be familiar, situated as it is on the road to Shark Bay, just after the turn-off from the Northern Highway. Hamelin Station is a 202 000 hectare rangelands property, strategically located adjacent to the Shark Bay World Heritage Area. It contributes to, and extends, all four of the Shark Bay World Heritage values: 'earth's history', 'ecological processes', 'natural beauty', and 'biological diversity'.[49] For Bush Heritage it presented as a unique new anchor property in the Mid West Rangelands Priority Landscape that would strengthen the links between Charles Darwin and Eurardy reserves, and the Gundawa Regional Conservation Assocation on the transitional vegetation zone called the 'mulga-eucalypt line' that straddles two major bioregions. In all these geographical capacities it would increase the representation and adequacy of some vegetation types in the NRS, and protect nationally significant populations of endemic reptiles and regionally significant populations of several threatened species, including the Hamelin skink, unique to the area. It would also build connectivity via other protected areas to

Eurardy Reserve, and help build protected native vegetation that stretches 200 kilometres from north to south.

As well as this, Hamelin Station has 40 kilometres of coastline on to Hamelin Pool on the eastern side of Shark Bay. And this is no ordinary coastline. Hamelin Pool 'has the largest and most diverse assemblage of living stromatolites in the world', according to Dr Erica Suosaari, a scientist from the US, who conducted a study of the stromatolites in 2014.[50] They live in the shallow water along the beach line of Hamelin Station, and look like rather spongy brown rock formations, bunched closely together. Stromatolites, the only living examples of earth's earliest life forms, are one of the organisms that produced oxygen, thereby helping to create the atmosphere on which life depends. Thus Hamelin Station is home to a remote and remarkable relatively undamaged environment that still nurtures the living source of life. Stromatolites are the only organisms that lived over 3 billion years ago for which we have fossil records, and their live presence in Hamelin Pool in great numbers evokes wonder for a place that is 'so raw and untouched and beautiful, unlike any other place on the planet'.[51] A thick bank of seagrass at the mouth of the bay impedes ready flushing of the water into the ocean so that the water maintains the particular salt intensity that supports the stromatolites. Another reason for the continued existence of this life source is that it has been largely left alone by other forces.

Bush Heritage Chief Executive Gerard O'Neill and Healthy Landscape Manager Greg Suosaari at Hamelin Station, WA

Photo: Ben Parkhurst

The owners of Hamelin, Brian and Mary Wake, managed the property for thirty-five years with a careful eye to conservation and for lease diversification, which saw them convert the shearing quarters into attractive accommodation for visitors, and develop a comfortable camping area, creating a welcome gateway to the Shark Bay region. When deciding to sell, they were keen to see the property in conservation hands and approached Bush Heritage as a possible buyer. They had done their homework and decided that Bush Heritage had the 'right philosophy'.[52] For Gerard O'Neill, it was a rare opportunity that would extend the World Heritage area of Shark Bay by up to 10 per cent through protection on its eastern side. With accommodation already available, the property would also provide a unique chance for Bush Heritage to host people in an easily accessible environment which could foster supporter and community conservation engagement. It could also be the location for a dynamic centre for a broad range of scientific research. O'Neill was quick to suggest Erica Suosaari as the first Science Fellow under its ten-year Science Plan. Her presence will encourage further research on the stromatolites, and bring the subject to a wider public as it engages with the enthusiastic and knowledgeable young scientist. And, as it turns out, she is married to Greg Suosaari, whose application to be the first manager of the property was successful. Both Luke Bayley and Dave Whitelaw had earlier made positive contact with the Traditional Owners, the Malgana people, who have indicated strong interest in working cooperatively with Bush Heritage to develop conservation and land management programs, and to share their long-term vision for their country with Bush Heritage.

In preparing to purchase the property, Bush Heritage developed a financial and fundraising strategy with a fundraising campaign target provisionally set at $4 million, to cover the cost of the property, initial set-up capital costs, the first three years of operating expenses, and a contribution to the Bush Heritage Capital Fund. Several significant donors were identified at this point, one of them being Ian Gust, who had helped buy the first Gondwana Link property. He and his wife had recently decided that, having made provision for their children, they wanted to make a substantial contribution to a conservation project. Instead of a general donation, Gust asked Bush Heritage if there was a particular project that might be especially interesting, and was delighted to support Hamelin.

Young had found the parrot in south-west Queensland, nowhere near the Fortescue mine, in typical outback desert country

Bush Heritage was keen to buy, and the Wakes to sell, and the deal should have been completed in 2014. The sale was not finalised for several months, however, because the WA government took its time signing off on the leasehold arrangements. Thus it was not until 2015 that Bush Heritage could announce the successful purchase of Hamelin Station, and begin to put in train some of the many threads of endeavour that will in the future make this property a showcase for both Bush Heritage and for the region. Already, Bush Heritage Board members, staff and supporters have gazed down into the shallow water from the lookout point on the jetty, sensing the existential pull of the weird-looking stromatolites that connect us to the origins of life. Work has begun on destocking the land of sheep, and rounding up thousands of feral goats, in expectation that after the next falls of rain, the country will begin to show the first signs of regeneration.

For O'Neill, Hamelin Station is a big project that will take up to ten years of hard work and partnership agreements before it begins to show the potential he believes it has.

O'Neill was also nursing a secret, even as he closed the deal on Hamelin Station. On the other side of the country, in south-west Queensland, things were moving on an equally exciting project which made it highly probable that Bush Heritage would soon become the owner of land that is the only known habitat of the night parrot. The night parrot is one of the most elusive and mysterious birds in the world. It is quite small, and predominantly yellowish-green in colour, with dark brown, black and yellow stippling, and a short tail. Because of its secretive nature, and the fact that it had only been sighted alive in 1912 and then once again in 1979, not much is known of its habits, except that it is a mainly ground-dwelling bird, only flying up from its spinifex habitat when startled, or when searching for food and water. Those characteristics that have made it 'invisible' in the landscape have seen it become the 'holy grail' for birdwatchers the world over, and although some believed it already extinct, there have been numerous attempts to discover its whereabouts. Hope that it would be found has been kept alive because of two quite serendipitous findings of dead parrots in recent times. In 1990 a carload of university ornithologists stopped to investigate a withered body lying on the side of the road, and identified it as the night parrot. In 2006 a grader driver saw another dead parrot hanging on a fence.

But the story of the night parrot's confirmed existence really only began when Fortescue Metals Group applied for approval for its $2 billion Cloud Break Mine project in the Pilbara region of Western Australia. As part of the process, it was required to implement a management plan to ensure that the mine would not adversely affect species' survival in the area. It commissioned a survey in April 2005 and, unfortunately for Fortescue, the two biologists employed to carry out the survey spotted three birds at dusk near Minga Well that they were sure were night parrots, although they were unable to photograph them. Their description was accepted by the Birdlife Australia Rarities

Researcher Stephen Murphy (left) and Bush Heritage science and research manager Jim Radford, trying to record night parrot calls

Source: Lyndon Mechielsen

Committee, however, and recorded as the only sighting of the parrot since 1979. Later, when Fortescue wanted to extend the mine, government Environment Protection and Biodiversity Conservation (EPBC) regulations were triggered, and Fortescue Metals Group was required to present a study showing why their project would not damage the night parrot. The Commonwealth engaged ornithologist Stephen Murphy to independently assess the advice that Fortescue handed to the Commonwealth.

From an early age, Stephen Murphy had wanted to be a ranger, dealing with wildlife, and was haunted by the disappearance of so many animal species. But he found the job unsatisfying, and returned to more academic wildlife research at the ANU, gaining a PhD in the ecology and conservation biology of palm cockatoos. Birds became his specialty because they seemed easy to study, and the night parrot interested him because he thought 'if we can find them, we could save them'.[53]

Murphy wrote his report for the government, and Tony Burke, Labor Minister for the Environment, indicated that Fortescue Metals could initiate the second stage of its mine on condition that it invest $1 million into research on the night parrot. At that stage Murphy was not interested in doing the research, since there was nothing to go on, and the bird could be anywhere. And then in May 2013 John Young, a wildlife cinematographer, took the first-ever photographs and video footage of a bird that was confirmed by experts as the night parrot. Young took on the research, choosing to keep the location secret because of poaching fears. But the project was too important for one person. He was advised to find someone who could help him with recording data. Young and Murphy knew each other, and Young finally confided in Murphy, although it took eighteen months of working closely together before he gave Murphy the location of the bird. Young had found the parrot in south-west Queensland, nowhere near the Fortescue mine, in typical outback desert country, where the red land greens up with young spinifex plants, the only groundcover across the plains and along the low rock outcrops that shelter small creeks in wet seasons. At the site, where they knew that the birds had been resident for at least two years, the two worked on ways to detect the presence of the birds, finding that they could rely only on their calls, since even after many long sessions in the field Murphy has only once seen an actual bird. They believe that the summer rains of the year before had triggered a breeding event, because they had picked up changes in the sound and frequency of the bird calls.

Finding a location in which the night parrot had resided for some time was the critical leap forward in deciding it was worthwhile to undertake more detailed research, even if the birds chose to leave at some point. Eighteen months into the project, Murphy and Young had developed a reliable and robust methodology for detecting the birds, based on their calls. But one day Murphy found thirty night parrot feathers on the ground. Cats! In that instant he knew that managing such an important project by themselves was too much of a responsibility. He could no longer be researcher and conservationist at the same time. It was time to bring in an organisation that had the capacity to take on the protection of the night parrot and conservation of its habitat. As Murphy remarked, he did not want to be remembered as 'the person who monitored night parrots into extinction'![54]

He knew that it was essential to move the cattle from the area, and that somehow this part of the very big cattle property needed to be quarantined as night parrot habitat. But how to do it? He had talked earlier with O'Neill about the project, discussing the possibilities of the kind of stewardship that Bush Heritage was pioneering in the Tasmanian Midlands, whereby landholders were paid to conserve their land. But the discovery of the bird feathers was the catalyst for Murphy to discuss the problem of cats with Rob Murphy (Bush Heritage Queensland executive manager, and no relation). Rob Murphy joined Bush Heritage in 2013, and his experience as regional director for Queensland Parks and Wildlife Service had included management of over 45 per cent of Queensland's protected areas within the vast and remote lands of western Queensland, so he was familiar with the region, and the threats that cats posed to native species. Stephen Murphy remembers the speed with which Rob Murphy jumped into action, sending two shooters down to the property. It was an automatic response that 'wasn't asking what was in it for Bush Heritage, just that there was a problem and something had to be done about it'.[55]

As discussions continued with both O'Neill and Rob Murphy about stewardship and how to work with the landholder, whose legitimate concern was his land and his cattle, Stephen Murphy reflected at length on which organisation had the capacity to handle the delicate situation in regard to protection of the night parrot. Murphy understood Bush Heritage's ethos, and had been impressed in the past with Bush Heritage's genuine engagement with different stakeholders; he did not know of any other organisation that approached conservation work with the same passion. Rob Murphy's response over the cat issue decided him that Bush Heritage was the most suitable organisation to entrust with the care of the parrot. Things moved quickly then, because of the urgency of the problem, and it was not long before he was sitting at a table with Rob Murphy and the landholder, discussing prices for the subdivision of 55 000 hectares of the property designated as night parrot habitat. Discussions continued until late 2015 as they worked out a deal in which it is hoped that the compensation for the loss of his land will encourage the farmer to support Bush Heritage's agenda in relation to protecting the night parrot's habitat. Bush Heritage took possession of the property in February 2016.

The first step is to remove the cattle, and erect fences to keep them out. Because the area has never suffered extensive fire damage, the thick groundcover has protected the birds. This means that cats are not as much a threat as was initially assumed, although control of the cats still presents the biggest challenge. Stephen Murphy has done several hundred hours of spotlighting in which time he has not seen a cat; indeed, the shooters commissioned by Rob Murphy spent two weeks on the property without sighting one. Since feral cats represent one of the biggest threats to native fauna throughout Australia, and are the cleverest at avoiding detection and capture, scientific research into their habits and areas of vulnerability is a top priority in Bush Heritage's new Science Plan, and has already begun with Tim Doherty's research at Charles Darwin Reserve into feral cat habits and habitat, and the pest review carried out on Goonderoo by Steve Jackson and Chris Dickman. But one important step, Stephen Murphy knew, was to stop the baiting of dingoes, to allow them to prey on the cats. This step would have to be negotiated with the owner of the land, who is now the closest neighbour, and still running a successful cattle business, and who regards dingoes as the greater threat. Another threat to the night parrot comes from human intervention, and Bush Heritage knows it will have to work closely with local people and authorities to regulate the flow of information about the bird, and visitation to the site, to protect the parrot against poachers and overenthusiastic birdwatchers.

At the time of writing, none of the big questions are resolved, but Bush Heritage is in the prime position to take on the protection of one of the most iconic and mysterious birds in the world. Already, donors are expressing their excitement with generous donations that have helped in securing the habitat.

It would seem that 2016 is shaping up to be Bush Heritage's busiest year yet, with two important new properties each moving Bush Heritage into new territory, and setting new benchmarks in the conservation world and for Bush Heritage. At the same time, work on the restoration of vegetation and the protection of animal and plant species continues on Bush Heritage properties and lands under Bush Heritage protection, and through partnerships with Aboriginal communities in every state and territory of Australia. Bush Heritage has come a long way from Bob Brown's first purchase of two bush blocks in the Liffey Valley. But his vision of an organisation that will remain into perpetuity still flames strongly for the Board and Bush Heritage employees and supporters. Together, over the past twenty-five years, they have provided the means, both financially and practically, for this to be a reality, and Bush Heritage will continue into the future as long as Australians want to protect and conserve their unique landscapes.

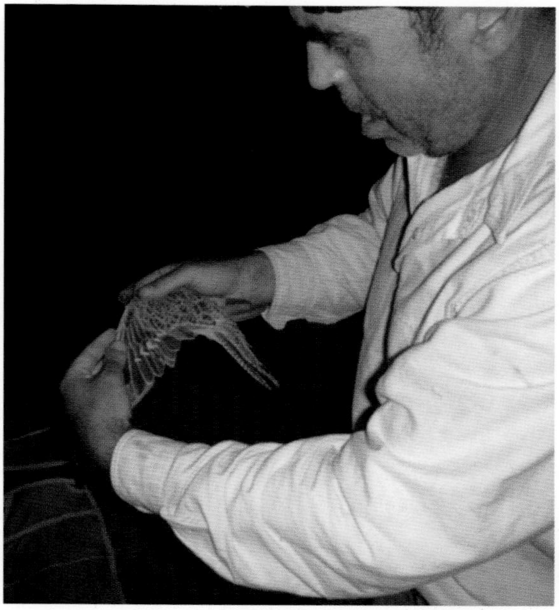

A moment to remember – a night parrot in the hand

Source: Rachel Barr

FOLLOWING SPREAD Arid coastal landscape belies the significance of the stromatolites in Hamelin Station, Shark Bay, WA

Source: Jiri Lochman/Lochman Transparencies

Epilogue

A year after my visit to Bob Brown, Peter and I are driving back to Oura Oura to climb Drys Bluff, the rocky ridge behind Bob's property on the Liffey River, now owned by Bush Heritage. No one is there when we arrive, but the sign welcoming trespassers is still on the bridge, and we take the track up to the cottage and beyond, up the slope to the wire fence and open gate on the boundary at the beginning of the walking track. A similar day to my last visit, sunny and cheerful, not overly hot; it is perfect for a walk up a hill. We have water, our Bush Heritage grab bag and lunch, which we will eat on the top of Drys Bluff.

The going is easy at first, and we have time to observe our surroundings, feel the softness of the thick leaf litter underfoot and smell the rich bush aromas. Gradually the incline becomes steeper, the vegetation sparser, although we are still in tall tree country. We come to a sheer rock wall, smooth granite, with long grasses like a fringe spilling over the top. We drink from the crystalline water that pours over the lip of the ledge, splashing noisily at our feet, and slipping down the mossy gutter into the gully. There is a fixed rope to grasp as we find the footholds in the rock face. On the top, the track is not so clear; perhaps most people stop at the water point and go no further. We traverse the hill, and angle up behind a granite outcrop with a flinty profile staring balefully down the hill. The track meanders fitfully, and we lose it as we clamber over low rocks and small plants. Trees sport large carbuncles and wonderfully textured bark coats, and each tree has an individual feature that merits a photograph. A magnificent bush, its branches laden with red berries, glows amidst the shades of green and brown.

Out in open space at last, below on the other side of the river we see a doll's house, a measure of how far we have come. But we can only guess where the top is, as a mist holds the secret. The barren slope of rock boulders will take us up to the final cliff face, we assume. We rock-hop up the slope, clumsily. Curious wallabies watch us with disdain or pity, we are not sure which, before bounding effortlessly away.

The uneven boulders funnel into a chimney-like gully, in which a narrow waterfall tumbles down the rock steps, crowded with small tough trees with leathery leaves, spikes and twisted branches, bent double in the jostling for place. Grasses, mosses, and small flowering plants thrive in the crevices, an ecosystem rich in variety and colour. But we don't stop, our goal is further up, and we are clambering now, hanging on to branches – will they hold us? – crossing and recrossing the trickling stream to find the easiest way up through the tangle. In the highest gully we reach the first of the ropes that Bob has written about, the ropes that the owners before him put in place to make the last 50 metres easier.[56] They do help, but it is a tough climb, as we let go of one rope and grab the next, with slippery footholds as purchase to pull ourselves up by the rope. The chimney narrows, until we squeeze through an aperture to arrive on the summit.

It is not really the summit. When was a summit so straightforward, so easy to attain at first sight? We cross the range, aiming for the highest point, along a narrow wallaby track disappearing beneath subalpine heaths and spongy bushes, following a meandering, sluggish stream flowing through coarse sand and moss-covered stones.

Paul Thomas on top of Drys Bluff, one of the highest points of the Great Western Tiers

Photo: Bob Brown

At last we are on the edge of the ridge on the highest point, in a space just under the sky. On three sides green trees stretch forever. We have to take Bob's word that you can see from Bass Strait in the north to forests near Hobart in the south, because the weather is closing in and soon we see not much further than 10 metres in front.[57] But low visibility can't stifle that sensation of infinite expansiveness and exhilarating aloneness, the result of achieving a summit after a hard climb.

In the car on the return journey to Launceston I have a sudden flash that our climb is emblematic of Bush Heritage's own journey.

We hunker down in thick bushes in the lee of the hill to eat lunch. The cold is intense, and we don't linger. On the downward path, we slide backwards down the chimney, hanging on to the ropes. Back on the sheltered side of the cliff, we thaw out a little, rock-hopping more slowly; descending is harder on the legs.

At last we gain the comparative ease of the dry forest. But the track has vanished, and we take the path of least resistance. We don't see the tree with the magnificent bole that Peter photographed on the way up, or the bush festooned with red berries. Never mind, we say, the only way is down! We search for familiar signs in bush that is denser than we remember. The first landmark that we come to is, disappointingly, the edge of the rock face, with the grass that hangs over it like a fringe. Peering over and along the cliff face, we realise now why the rope was where it was: it is, in fact, the only way to ascend or descend, unless we fancy a slip-and-slide over a 20-metre cliff. No, we don't. We realise we have come too far across the face of the hillside and retreat, our interest in finding a landmark heightened by the knowledge that there is actually only one way down.

At last we see it: our Easter Island man staring down into the valley. The only big crag in the vicinity, his profile is unmistakeable. We join the track again, but not before climbing onto his head to view the valley below still patchworked in sunlight and cloud shadows.

After that, descending to the lower slopes through the tall trees with their underlay of leaf litter and small ferns is quite pleasant. But we are beginning to feel the strain in our legs, not surprising since we have done 1000 metres up and back in four hours. The last part, on relatively flat ground, is the hardest. My legs are aching and I struggle to put one foot in front of the other. I feel I won't make it, but Peter takes my backpack on the final stretch, and we reach the road, exhausted but triumphant.

In the car on the return journey to Launceston I have a sudden flash that our climb is emblematic of Bush Heritage's own journey. I brush the thought away as far-fetched. But I can't help feeling that the day has in some way clarified some of the confusion I have experienced while working my way through the twenty-five years of Bush Heritage's life. As the years accumulated, the material became more diffuse, more difficult to organise. So much was happening in so many directions, some events more sharply defined, while others by contrast receded into the background. I had begun to notice, as I was writing, irregularities in tenses, and an increase in the use of the present tense. I had wanted to change back into the past tense, but stopped myself. And it dawns on me now, that there is a watershed moment in the 'history', a ridge. On one side history peels back and down into the hinterland, distinguishable features against the more muted green of references and interpretation; on the other side the future stretches out, also signposted by salient features, but with a longer vista that disappears into a hazy horizon.

I reflect on the problems of writing a history that does not have a conclusion. A history with a finishing date means you can recount verifiable actions by reference to sources, and it permits an authorial dimension, since history is informed by the present. But the present gives a different perspective to the story, and it is less easy to discern which events and actions will define the moment; there is only conjecture based on past events and what is happening now.

In Bush Heritage's case the difficulty is multiplied. By its own declared mission there will never be a conclusion. Bush Heritage aims to be there in perpetuity, protecting Australia's landscapes, whatever it takes. The past will always colour the present, and together they will go towards creating the future, but we cannot guess how known and unknown forces will bear on the trajectory Bush Heritage has set itself. Later, when I tell Bob that I came back to climb Drys Bluff, he comments that it is one of the hardest day walks in Tasmania, and once again I have that strange flash of recognising in our hike up and down this forbidding but beautiful track an allegorical connection to Bush Heritage's own story.

I let the landscapes we have seen flow across my mind, a moving, vivid kaleidoscope of colour and shapes and living things.

Flying home to Melbourne, I think about the tall tree forests of Tasmania, where the Bush Heritage story began. There is something awe-inspiring in the straight tall trunks that stretch forever above you. But there is a uniformity in these forests that I did not find elsewhere in our travels around Australia, as we wandered through graceful salmon gum woodlands, and fell in love with the York gums, the gimlet, the *wubeninsia* eucalypt, and other box gums, the mulga in its many forms, particularly the silver and the red mulgas; the different types of mallee, the gidgee, the myall, the brigalow and the hundred different varieties of acacia and callistemon and fig that we have seen across the continent. A mind-numbing, bewildering array of trees, all beautiful and all different. I ask Peter which is his favourite, and we spend the rest of the flight discussing the various attributes of each type of tree, until I decide on the myall, with its flowing, shapely limbs that stretch horizontally, each sporting deep green feathery foliage in clouds on the end of slender branches, while Peter goes for the silver mulga, a wonderful bushy shrub that glitters in the sunlight, a solid companion on red sands in spinifex country. For a country with very poor soils, Australia has produced a staggering number of different tree types, far more than are found in other parts of the world. But then trees are only the most noticeable manifestation of the variety and range of life here: that same dazzlingly complex diversity is in the grasses, the plants, in the animals and birds, in the thousands of species of insects in the natural world around us that we Australians regard with a certain amount of complacency as a birthright.

In the course of my research, I have learnt the reason for this wondrous variety of plant life: the separation from the rest of the world's biota some 40–50 million years ago first influenced diversity of life and form. And then the long history of isolation and complex environmental changes across the Australian continent imposed its own massive internal variation, during times when the land was mostly covered by shallow seas and at others was subjected to harsh and cold conditions, or intense periods of heat and drought. All these different eras were opportunities for evolution's slow process of enforcing changes, changes that were repeated in endless cycles of variation, forcing genetic variability to its limits, and resulting in the staggering diversity that we see today. This process, as I understand, created something like a production line for diversity, a species pump, where small populations of living things have been isolated and exposed to intense selective pressure. If they survived and adapted, they were, when reunited with their ancestors, sufficiently changed as to be unrecognisable, a new species in fact. In my mind I let the landscapes we have seen flow across my mind, a moving, vivid kaleidoscope of colour and shapes and living things, an intensity of experience that includes the people who have told me about themselves, have shown me their land and the work they do to care for the life that we risk taking for granted.

Thinking about their stories makes me worry for the future, however. Almost everyone I have met who has spent a major part of their life or their income in conservation of the natural bush first learnt their passion in childhood. Some event or some experience when they were quite young taught them to love nature and to want to protect it. But as the natural landscape recedes from us, and camping and bushwalking, or just mucking around in the long grass on a vacant block, or collecting tadpoles in a glass jar from a small pond, are replaced by other more sedentary and button-pushing pursuits, I wonder about the next generation of carers for the bush, and the next, and the next. Will they find their way to the wonder and excitement of discovery of the natural world, and make protecting it part of their life? And without those people prepared to make a stand against exploitation of what is left of our bushland, and keen to live in remote places to watch over it, there is every chance it will disappear without anyone even noticing. As Simon Barnes, in one of his last 'Wild Notebooks' for the *Times*, in June 2014, poignantly regrets:

> Where there are fewer insects there are fewer insect-eaters, and where there are fewer insect-eaters there are fewer insect-eater-eaters. The whole ecosystem is changing. That worries me very much and not just because we are losing bio-abundance at a truly alarming rate. It's something to do with expectation. Conservation begins with a sense of loss, an urge to try to bring the world back to the state we knew in our childhood. The founder-members of the conservation movement began their work in a world already grossly depleted ... We all find ourselves satisfied with less. As a result wildlife gets even more marginalised and we humans are ever more alienated from the planet we live on, complaining of a thousand urban anxieties while the solution to those problems withers and dies all around us.[58]

A couple of days later, I am walking with my small granddaughter through a park in inner Melbourne, still not quite able to stop thinking about this problem, and how it might affect her wellbeing, and her children's wellbeing. Selfishly, that is about as far as I can project into the future. We are on our way, Wanda and I, to the swings and slides, but for some perverse reason I have chosen the roundabout route, through an avenue of oak trees along a lightly gravelled path. It occurs to me that the nineteenth-century town planners of this city were generous in their allocation of park spaces and that this must have been an early park because the trees are quite old. I know they are old because each tree we pass has some kind of hollow, either in its trunk, or at the point where branches join the trunk.

'Look, Wanda,' I tell her, 'that's where animals live, or a bird nests.' She is just two years old, so I don't really expect her to understand. But she knows about owls. 'Too-wit too-woo!' she exclaims, and points to the hollow up in the tree, above her reach. I lift her up and we peer in. No bird, not in this hollow, nor in the next. I warm to my theme and we talk about possums and lizards, and even spiders and beetles that might live in the hollows, and she insists on inspecting every hollow. But we don't find even the trace of a dead spider, and I am becoming tired of lifting her up to every hollow. I walk faster to reach the swings for a rest.

I wonder about the next generation of carers for the bush, and the next, and the next. Will they find their way to the wonder and excitement of discovery of the natural world, and make protecting it part of their life?

As I push her 'higher, higher' on the swing, birds fly overhead, screeching. 'Parrots!' she cries. She already knows about parrots and magpies and kookaburras. And imperiously demands that I stop the swing. 'Find parrots,' she says, and sets off at a two-year-old run towards a big old Moreton Bay fig tree not far from the playground, and from which we can hear the busy gossiping of many birds. As we arrive at the foot of the tree, the birds take off. They are lorikeets, and we catch flashes of iridescent green, blue, red and yellow before they disappear. Wanda is disappointed. It looks as if we've disturbed them and they won't come back. But no.

There are two left. We hear their chatter before we see them. As we search for the source of the noise with our eyes, one lorikeet pops out of a hollow in the tree, and then the other. They chortle together for a bit, hopping along the branch. One swaggers back into the hollow, poking its head out for a last admonishment to the other, which quickly follows its companion back into the hollow, tail feathers waggling as it disappears from sight.

I look down at Wanda. 'Hollow!' she breathes, her whole body straining upwards to catch another glimpse of the birds, her eyes fixed on the hollow in the tree. Her face is alive with discovery; she has understood about hollows in trees. It is a start.

Notes

Prologue

1 Watson, Don, *The Bush: Travels in the heart of Australia*, Penguin Books, Melbourne, 2014.

PART 1

Bob Brown's vision

1 Paul Smith, interview, 2014.

2 Bob Brown, *Optimism: Reflections on a life of action*, Hardie Grant Books, Melbourne, 2014, p 1.

3 Brown, *Optimism*, p 2.

4 *The Last Wild River*, film by Paul Smith and Amanda Stark, Tas (predates 1979) <trove.nla.gov.au/work/21844695>

5 Phillip Toyne, interview, 2014.

6 Bob Brown, interview, 2015.

7 Brown, 2015.

8 Goldman Environmental Prize <www.goldmanprize.org/about/>

9 Brown, 2015.

10 Brown, 2015.

11 Judy Henderson, interview, 2014.

12 Bob Brown, 'The quest for Drys Bluff: A view from Liffey', *Habitat Australia*, vol 18, no 4, August 1990.

13 Brown, 2015.

14 Peg Putt, interview, 2014.

15 Brown, 2015.

16 Verna Simpson, interview, 2014.

17 Simpson, 2014.

18 Reported in an article profiling early supporters of Bush Heritage, *Bush Heritage News*, Spring 2011.

19 Putt, 2014.

20 Sally Port, interview, 2014.

21 Brown, 2015.

22 Bob Brown, President's Report, Annual General Meeting, Australian Bush Heritage Fund, 1995.

23 Greg Blake, interview, 2014.

24 Port, 2014.

25 Simpson, 2014.

26 Port, 2014.

27 Minutes, Australian Bush Heritage Fund meeting, November 1992.

28 Penny King, interview, 2014.

29 Jeremy Hobbs, interview, 2015.

30 King, 2014.

31 *Australian Bush Heritage Fund News*, October 1992.

32 *Australian Bush Heritage Fund News*, October 1992.

33 *Grammardog Guide to Walden by Henry David Thoreau*, Grammardog LLC, 2007, p 25.

34 Simon Schama, *Landscape and Memory*, Fontana Press, London, 1996, p 6.

35 John Muir, quoted in Schama, *Landscape and Memory*, p 7.

36 ED Hoben, 'A national pleasure ground. The Kuring-gai Chase', *Sydney Morning Herald*, 14 April 1900, p 12 <trove.nla.gov.au/ndp/del/article/14305781>

37 Donella H Meadows, Dennis L Meadows, Jørgen Randers and William W Behrens III, *Limits to Growth*, commissioned by Club of Rome, Universe Books, 1972.

38 Margaret Graham, interview, 2014.

39 Michael Williams, *Bush Heritage News*, Summer 2010, p 11.

40 World Wide Fund for Nature <www.wwf.org.au/about_us/wwf_australias_missions_and_goals/>

41 J Ros Garnet, 'The National Parks and the FNCV', *Victorian Naturalist*, vol 97, no 3, 1980, pp 130–34.

42 Libby Robin, *Defending the Little Desert: The rise of ecological consciousness in Australia*, Melbourne University Press, 1998.

43 Monica Dux, 'Growing up on troubled ground', *Age*, 13 December 2014.

44 Bryan Keon-Cohen, *A Mabo Memoir: Islan kustom to native title*, Zemvic Press, Melbourne, 2013.

45 Whitlam government environmental policy <whitlamdismissal.com/1972/11/13/whitlam-1972-election-policy-speech.html>

46 ibid.

47 Max Bourke, interview, 2015.

48 Richard Hines, communications manager, Greening Australia. Interview, 2015.

49 Graham Richardson, *Whatever it Takes*, Bantam, Sydney, 1994, p 214.

50 RJL Hawke, *Our Country Our Future: Statement on the environment*, July 1989 <trove.nla.gov.au/work/16690946>.

51 'Forestry peace deal: The end of another chapter', Andrew Fisher, *The Drum*, ABC Television, 20 October 2010.

52 Brown, 2015.

53 Karen Alexander, interview, 2014.

54 Sharon Beder, 'Activism versus negotiation: Strategies for the environment movement', *Social Alternatives*, vol 10, no 4, December 1991, pp 53–56.

55 'Who are you? Martin Copley', ABC Local Radio Perth, 21 March 2013 <mpegmedia.abc.net.au/local/perth/201303/r1090292_13049116.mp3>

56 'Wamsley walks away from Earth Sanctuaries', Reporter: Prue Adams, *Landline*, ABC Television, 27 March 2005.

57 Bob Brown, quoted in James Norman, *Bob Brown: Gentle revolutionary*, Allen & Unwin, Sydney, 2004, p 74.

58 Margaret Atkinson, interview, 2014.

59 Robert Rosen, interview, 2014.

60 This book uses 'Allens'.

61 Greg Blake, interview, 2014.

62 Jamie Kirkpatrick, interview, 2014.

63 EO Wilson (ed), *Biodiversity*, National Academy Press, Washington, DC, 1988.

64 Wilson (ed), *Biodiversity*, p v.

65 David Ehrenfeld, 'Why put a value on biodiversity?', in Wilson (ed), *Biodiversity*, p 212.

66 Rio Earth Summit, 1992 <www.un.org/geninfo/bp/envirp2.html>

67 PJ Keating, 'Statement on the Environment', December 1992 <www.environment.gov.au/land/nrs/about-nrs/history>.

68 Henderson, 2014.

69 Janice Bird, interview, 2014.

70 Bird, 2014.

71 Bird, 2014.

72 Margaret Graham, interview, 2014.

73 Brown, 2015.

74 Peter Christoff, quoted in Susan Wilson, 'Karen Alexander', *Wild Magazine*, July–August 2010, pp 18–19.

75 Mara Bún, interview, 2014.

76 Paul Sattler, interview, 2014.

77 Australian Government, Department of the Environment, 'History of the National Reserve System' <www.environment.gov.au/topics/land/nrs/science/scientific-framework>

78 Greg Blake, *Managing the Matrix: A plain language manual to help you manage your bush*, Reserve Design and Management, Dunalley, Tas, 2003.

79 Blake, 2014.

80 Blake, 2014.

81 Bird, 2014.

82 Denise Krake, a volunteer, living at Brogo. Interview, 2015.

83 Letter from J O'Halloran to Beth Schultz, 14 February1995, Margaret Robertson private files.

84 Margaret Robertson, interview, 2014.

85 The personal story of farmer William Harvey – part of 'Creating the Farms' display, Kodja Place, Kojonup, WA.

86 'Bush buy back beats abuse', *Great Southern Herald*, 19 March 1997, p 13.

87 'Bush buy back beats abuse'.

88 Blake, 2014.

89 Keith Bradby, interview, 2014.

90 Blake, 2014.

91 Report on the purchase of the Erith Island lease, *Australian Bush Heritage Fund Annual Report, 1996–97.*

92 James Button, 'Wild idyll', *Age*, 21 March 1997.

93 Brown, 2015.

94 Alexander, 2014.

95 Doug Humann, interview, 2014.

96 Doug Humann, 'A personal journey to innovation', in Penelope Figgis, James Fitzsimons and Jason Irving (eds), *Innovation for 21st Century Conservation*, Australian Committee for IUCN, Sydney, 2012, pp 16–23.

97 *Australian Bush Heritage Fund Annual Report, 1996–97.*

98 Brown, 2015.

99 Brown, 2015.

100 Katrina Blake, interview, 2014.

101 Peter Crosier, Wentworth Group of Concerned Scientists. Interview, 2015.

102 Humann, 2014.

103 Nathan Males, interview, 2014.

104 Rod Fensham, ecologist working at Queensland's Herbarium and at the University of Queensland. Interview, 2014.

105 Hugo Spooner, interview, 2014.

106 Spooner, 2014.

107 Brown, 2015.

108 *Environment Protection and Biodiversity Conservation Act 1999* <www.comlaw.gov.au/Details/C2015C00422>

109 *Australian Bush Heritage Fund Annual Report 1999.*

110 Rosen, 2014.

111 Louise Gilfedder, interview, 2014.

112 Gilfedder, 2004.

113 Males, 2014.

114 Bird, 2014.

115 G Blake, interview, 2014.

PART 2

Making the vision a reality

1 Greg Blake, interview, 2014.

2 R and V Routley, *The Fight for the Forests: The takeover* of *Australian forests for pines, wood chips and intensive forestry*, Research School of Social Sciences, ANU, Canberra, 1973.

3 Margaret Robertson, interview, 2014.

4 Karen Alexander, interview, 2014.

5 Doug Humann, interview, 2014.

6 Both later indicated to Humann that they believed they had been wrong at the time. Humann, 2014.

7 Humann, 2014.

8 Brian and Diana Snape, interview, 2014.

9 Alexander, 2014.

10 Libby Smith, 'Carnarvon Station: A history of European settlement since 1863', Bush Heritage Australia, 2003 <www.bushheritage.org.au/places-we-protect/queensland/carnarvon>

11 Interview with anonymous donor, 2014.

12 Alexander, 2014.

13 Mara Bún, interview, 2014.

14 Alexander, 2014.

15 Phillip Toyne, interview, 2014.

16 Toyne, 2014.

17 'Canberra close up: Phillip Toyne', interview with Alex Sloan, ABC Local Radio Canberra, 21 March 2012 <www.abc.net.au/local/stories/2012/03/21/3460338.htm>

18 'Canberra close up'.

19 'Canberra close up'.

20 Phillip Toyne, *Growing up the Country*, Penguin Books, Melbourne, 1987, p 7.

21 John Bradley with Yanyuwa families, *Singing Saltwater Country: Journey to the songlines of Carpentaria*, Allen & Unwin, Sydney, 2010.

22 Toyne, *Growing up the Country*, p 8.

23 Toyne, 2014.

24 Humann, 2014.

25 Nicholas Brown and Susan Boden, *A Way Through: The life of Rick Farley*, NewSouth Publishing, Sydney, 2012, p 169.

26 Phillip Toyne, quoted in Brown and Boden, *A Way Through*, p 172.

27 Phillip Toyne, quoted in Brown and Boden, *A Way Through*, p 169.

28 Jane Elix, quoted in Brown and Boden, *A Way Through*, p 168.

29 Humann, 2014.

30 'Carnarvon Station Reserve – Launched', *Bush Heritage News*, Spring 2001.

31 'Carnarvon Station Reserve – Launched'.

32 Kate Fitzherbert, interview, 2014.

33 Chris Darwin, interview, 2014; Steve Morton, interview, 2014.

34 Bill Birchard, *Nature's Keepers: The remarkable story of how The Nature Conservancy became the largest environmental organization in the world*, Jossey Bass, San Francisco, 2005, p 113.

35 Kent Wommack, interview, 2014.

36 Wommack, 2014.

37 Wommack, 2014.

38 Birchard, *Nature's Keepers*, p 105.

39 Fitzherbert, 2014.

40 Recounted by Fitzherbert, 2014.

41 Kent Wommack, 'Confidential Report' on Bush Heritage to TNC Board, June 2001.

42 Wommack, 'Confidential Report'.

43 Doug Humann, describing US donors' amazement at the price of pastoral land in Australia. Interview, 2014.

44 Humann, 2014.

45 Stuart Cowell, interview, 2014.

46 Nathan Males, interview, 2014.

47 Males, 2014.

48 Louise Gilfedder, Jamie Kirkpatrick, Hans Wapstra and Annie Wapstra (eds), *The Nature of the Midlands*, Midlands Bushweb: Northern Midlands Council, Longford, Tas, 2003.

49 K Mokany, D Friend, J Kirkpatrick and L Gilfedder, 'Managing Tasmanian Native Pastures: A Technical Guide for Graziers', Tasmanian Institute of Agricultural Research, University of Tasmania, 2006 <dpipwe.tas.gov.au/conservation/publications-forms-and-permits/publications/managing-tasmanian-native-pastures>

50 Julian von Bibra, 'Midlands Conservation Fund: Managing natural capital', Prospectus, Bush Heritage, 2015, p 3.

51 Von Bibra, 'Midlands Conservation Fund'.

52 Bob Brown, interview, 2015.

53 Brown, 2015.

54 Keith Bradby, 'Gondwana Link: Process or plan, movement or organisation?', in Penelope Figgis, James Fitzsimons and Jason Irving (eds), *Innovation for 21st Century Conservation*, Australian Committee for IUCN, Sydney, 2012, p 100.

55 Colin Cameron, in *The West Australian*, 1 April 1959, quoted in Kaye Vaux, 'The view east from Ongerup', in *The Bush Comes to the City: Fitzgerald Biosphere Project* (proceedings of a seminar held at Murdoch University, Perth, 25 September 1987), Fitzgerald Biosphere Project, 1989, p 4.

56 Angela Sanders, interview, 2015.

57 Sanders, 2015.

58 Keith Bradby, WA conservationist. Interview, 2014.

59 Bradby, 'Gondwana Link', p 102.

60 Kate Fitzherbert's recollection. Interview, 2014.

61 Bradby, 2014.

62 Cowell, 2014.

63 Cowell, 2014.

64 Lance H Gunderson, CS Holling and Stephen S Light (eds), *Barriers and Bridges to the Renewal of Ecosystems and Institutions*, Columbia University Press, New York, 1995. The first in the series was: CS Holling, *Adaptive Environmental Assessment and Management*, John Wiley & Sons, Chichester, UK, 1978.

65 Humann, 2014.

66 Morton, 2014.

67 Morton, 2014.

68 Cowell, 2014.

69 Annette Stewart, interview, 2014.

70 Robert Rosen, interview, 2014.

71 Robert Rosen, 'ALES, Version 2', 2 July 1999, Bush Heritage Archives.

PART 3

Expanding Australia-wide

1 Libby Smith, 'Carnarvon Station: A history of European settlement since 1863', Bush Heritage Australia, 2003, pp 28–32 <www.bushheritage.org.au/places-we-protect/queensland/carnarvon>.

2 John Cianchi, 'I talked to my tree and my tree talked back: Radical environmental activists and their relationships with nature', PhD thesis, University of Tasmania, 2013.

3 Hugo Spooner, interview, 2014.

4 Peter Menkhorst and Frank Knight, *A Field Guide to the Manmmals of Australia*, Oxford Univeristy Press, Melbourne, 2001.

5 'Feral cats rewrite the Australian story', *Background Briefing*, Reporter: Gregg Borschmann, ABC Radio RN, 12 October 2014 <www.abc.net.au/radionational/programs/backgroundbriefing/feral-cats-re-write-the-australian-story/5802204#transcript>.

6 Steve Heggie, interview, 2015.

7 Heggie, 2015.

8 Spooner, 2014.

9 Charles Darwin discussing his religious belief. Nora Barlow (ed), *Autobiography of Charles Darwin 1809–1882*, Collins, London, 1958, p 95 <darwin-online.org.uk/content/t?viewtype=text&itemID=F1497&pageseq=1>.

10 Barry Traill, interview, 2015.

11 Chris Darwin, interview, 2015.

12 Darwin, 2015.

13 Darwin, 2015.

14 Darwin, 2015.

15 Charlie Nicholson, interview, 2014.

16 Graeme L Worboys and Brendan Mackey, 'Connectivity conservation initiatives', in James Fitzsimons, Ian Pulsford and Geoff Westcott (eds), *Linking Australia's Landscapes: Lessons and opportunities from large-scale conservation networks*, CSIRO Publishing, Melbourne, 2013, p 7.

17 Bruce Chatwin, *The Songlines*, Penguin Books, New York, 1988.

18 John Bradley, *Singing Saltwater Country: Journey to the songlines of Carpentaria*, Allen & Unwin, Sydney, 2010.

19 Bryan Keon-Cohen, *A Mabo Memoir: Islan kustom to native title*, Zemvic Press, Melbourne, 2013.

20 Keon-Cohen, *A Mabo Memoir*.

21 Bill Gammage, *The Biggest Estate on Earth: How Aborigines made Australia*, Allen & Unwin, Sydney, 2011. Bruce Pascoe, *Dark Emu Black Seeds: Agriculture or accident?*, Magabala Books, Broome, WA, 2014.

22 R Dawson, *The Present State of Australia* [1830], Alburgh, UK, 1987, pp 108–09, cited in Gammage, *The Biggest Estate on Earth*, p 14.

23 Gammage, *The Biggest Estate on Earth*, p 61.

24 Sarah Eccles, interview, 2014.

25 Stuart Cowell, interview, 2014.

26 Cowell, 2014.

27 Emilie Ens, 'Conducting two-way ecological research', in Jon Altman and Seán Kerins (eds), *People on Country*, Federation Press, Sydney, 2012, p 52.

28 Eccles, 2014.

29 Email, Kent Wommack to Stephanie Speakman, November 2003.

30 Guy Fitzhardinge, interview, 2014.

31 The Pew Charitable Trusts <www.pewtrusts.org/en>.

32 John Woinarski, Barry Traill and Carol Booth, 'The Modern Outback: Nature, people and the future of remote Australia', Pew Charitable Trusts, 2014, pp iii, 2 <www.pewtrusts.org/~/media/assets/2014/10/themodernoutbackforweb.pdf>.

33 Woinarski, Traill and Booth, 'The Modern Outback', p 3.

34 Doug Humann, private papers.

35 Barry Traill, 'Populate or perish', *Australian Geographic*, vol 124, January–February 2015, p 39.

36 Traill, 'Populate or perish'.

37 Roewen Wishart, Great Southern Arc fundraising campaign notes, 2005.

38 'Elizabeth George, supporter', *Bush Heritage News*, Spring 2011, p 9.

39 John Bailey, *Mr Stuart's Track: The forgotten life of Australia's greatest explorer*, Picador, Sydney, 2007.

40 'Generous donation saves Australian bushland', *The 7.30 Report*, Reporter: Kerry Brewster, ABC Television, 15 May 2008.

41 Kosciuszko2Coast <k2c.org.au/PeterBarrer>.

42 Kosciuszko2Coast <k2c.org.au/K2C>.

43 Ian Pulsford, interview, 2015.

44 Pulsford, 2015.

45 Paul Foreman, interview, 2014.

46 Katrina Blake, interview, 2014.

47 Blake, 2014.

48 Annette Stewart, interview, 2014.

49 Stewart, 2014.

50 Stewart, 2014.

51 Phillip Toyne, *Bush Heritage News*, Autumn 2007.

52 Humann, 2014.

53 Nicola Markus, interview, 2014.

PART 4

Preparing for perpetuity

1 Stuart Cowell, interview, 2014.

2 Louise Sylvan, interview, 2014.

3 Nicola Markus, *On Our Watch: The race to save Australia's environment*, Melbourne University Press, 2009.

4 Nicola Markus, interview, 2015.

5 Dave Whitelaw, interview, 2014.

6 Davis McCaughey, 'An ignorant man asking questions', Annual Muresk Lecture, Western Australian University of Technology, 1984.

7 EO Wilson (ed), *Biodiversity*, National Academy Press, Washington, DC, 1988.

8 David Ehrenfield, 'Why put a value on biodiversity?', in Wilson (ed), *Biodiversity*, p 215.

9 Max Tischler, former ecologist with Bush Heritage, quoting his friend and colleague, Andrew Harper. Interview, 2015.

10 Steve Heggie, interview, 2015.

11 Darren Larcombe, interview, 2014.

12 Glen Norris, interview, 2014.

13 Karen Harrland, *Spinifex Baby*, Finch Publishing, Melbourne, 2014.

14 Sandy Gilmore, Bush Heritage Blog, 23 October 2014 <blog.bushheritage.org.au/blog/author/60/sandy-gilmore>

15 Adam Kerezsy, *Desert Fishing Lessons: Adventures in Australia's rivers*, UWA Publishing, Perth, 2011. Adam Kerezsy, interview, 2014.

16 Renee Rossini, Bush Heritage Blog, 31 March 2015 <blog.bushheritage.org.au/blog/author/129/renee-rossini>

17 Rossini, Bush Heritage Blog.

18 Renee Rossini, interview, 2014.

19 Rossini, 2014.

20 Tim Low, *Where Song Began: Australia's birds and how they changed the world*, Penguin Books, Melbourne, 2014, p 20.

21 Vanessa Westcott, Bush Heritage Blog, 21 March 2014 <blog.bushheritage.org.au/blog/article/vanessas-first-post-as-the-new-regional-ecologist-wa-rangelands>

22 Phillip Toyne, interview, 2014.

23 Brendan Mackey, Sandy Gilmore, Hugh Pringle, Paul Foreman, Linda Van Bommel, Sandra Berry and Murray Haseler, 'BioPrEP: A regional, process-based approach for assessment of land with high conservation value for Bush Heritage Australia', *Ecological Management and Restoration*, vol 11, no 1, April 2010, pp 51–60.

24 Doug Humann, interview, 2014.

25 Gerard O'Neill, interview, 2015.

26 O'Neill, 2015.

27 Seán Kerins, interview, 2015.

28 Jack Green, interview, 2015.

29 Shaun Ansell, interview, 2015.

30 Phillippa Walsh, interview, 2015.

31 O'Neill, 2015.

32 O'Neill, 2015.

33 Bush Heritage/Adnyamathanha partnership <www.bushheritage.org.au/what_we_do/aboriginal-partnerships/adnyamathanha>

34 Mackey and others, 'BioPrEP'.

35 Luke Bayley, interview, 2014.

36 Peter Saunders, interview, 2014.

37 Bill Thompson, interview, 2015.

38 Jim Radford, 'The magic of Naree: Wetlands in the outback,' *Bush Heritage Newsletter*, Summer 2012.

39 Stacey Irving, 'A love for life on Naree', *Bush Heritage Newsletter*, Summer 2012.

40 'Bush Blitz', *Living Black*, NITV (National Indigenous Television), 21 May 2013 <www.sbs.com.au/ondemand/video/30904387777/living-black-bush-blitz>

41 Ongoing discussions with Bush Heritage insurers preclude further disclosure of the facts surrounding Pieterse's death.

42 'Launching Bush Heritage's Ten Year Science Plan', literature accompanying the launch at Scottsdale, March 2015. Information is also available at <www.environment.gov.au/biodiversity/conservation/hotspots>

43 Allana Brown, interview, 2015.

44 Sue Akers, interview, 2015.

45 Jody Gunn, interview, 2015.

46 Reported by Gerard O'Neill. Interview, 2015.

47 Kerezsy, 2014.

48 Steve Heggie, interview, 2015.

49 Shark Bay website <www.sharkbay.org.au/world-heritage/wha_values.aspx>

50 Erica Suosaari, interview, 2014.

51 Erica Suosaari, video presentation.

52 Mary Wake, video presentation, interview by Cineport Media, 2015.

53 Stephen Murphy, interview, 2015.

54 Murphy, 2015.

55 Murphy, 2015.

56 Bob Brown, 'Rhapsody of Oura Oura', *Ampersand Magazine*, no. 4, Spring 2011, pp 37–40.

57 Brown, 2011.

58 Simon Barnes, 'Without a sky full of insects there's no sky full of birds', Wild Notebook, *The Times*, 7 June 2014.

Bibliography

Books and chapters

Abram, David, *The Spell of the Sensuous: Perception and language in a more-than-human world*, Vintage, New York, 1997.

Altman, Jon, and Seán Kerins (eds), *People on Country: Vital landscapes, indigenous futures*, Federation Press, Sydney, 2012.

Bailey, John, *Mr Stuart's Track: The forgotten life of Australia's greatest explorer*, Picador, Sydney, 2007.

Birchard, Bill, *Nature's Keepers: The remarkable story of how The Nature Conservancy became the largest environmental organization in the world*, Jossey Bass, San Francisco, 2005.

Blake, Gregory, *Managing the Matrix: A plain language manual to help you manage your bush*, Reserve Design and Management, Dunalley, Tas, 2003.

Bolton, Geoffrey, *Spoils and Spoilers: Australians make their environment 1788–1980*, Australian Experience No 2, George Allen & Unwin, Sydney, 1981.

Bourke, Max, 'Why we need Rick Farley now more than ever', in Figgis, Fitzsimons and Irving (eds), *Innovation for 21st Century Conservation*, pp 30–37.

Bradby, Keith, *A Rationale for Respect, in Diversity or Dust: A review of the impact of agricultural land clearance programmes in south-west Australia*, Australian Conservation Foundation, Melbourne, 1984.

—— 'Gondwana Link: Process or plan, movement or organisation', in Figgis, Fitzsimons and Irving (eds), *Innovation for 21st Century Conservation*, pp 100–07.

Bradley, John, with Yanyuwa families, *Singing Saltwater Country: Journey to the songlines of Carpentaria*, Allen & Unwin, Sydney, 2010.

Brown, Bob, *Memo for a Saner World*, Penguin Books, Melbourne, 2004.

—— *Optimism: Reflections on a life of action*, Hardie Grant Books, Melbourne, 2014.

Brown, Nicholas, and Susan Boden, *A Way Through: The life of Rick Farley*, NewSouth Publishing, Sydney, 2012.

Carson, Rachel, *Silent Spring*, Houghton Mifflin, Boston, 1962.

Chatwin, Bruce, *The Songlines,* Penguin Books, New York, 1988.

Diamond, Jared, *Collapse: How societies choose to fail or survive*, Penguin Books, New York, 2005.

Ehrenfeld, David, 'Why put a value on biodiversity?', in Wilson (ed), *Biodiversity*, pp 212–16.

Ehrlich, Paul, *The Population Bomb*, Ballantyne Books, New York, 1968.

Ens, Emilie, 'Conducting two-way ecological research', in Altman and Kerins (eds), *People on Country.*

Figgis, Penelope, James Fitzsimons and Jason Irving (eds), *Innovation for 21st Century Conservation*, Australian Committee for IUCN, Sydney, 2012.

Fitzsimons, James, Ian Pulsford and Geoff Westcott (eds), *Linking Australia's Landscapes: Lessons and opportunities from large-scale conservation networks*, CSIRO Publishing, Melbourne, 2013.

Flannery, Tim, *The Future Eaters: Ecological history of the Australasian lands and people*, New Holland Publishers, Sydney, 2006.

Gammage, Bill, *The Biggest Estate on Earth: How Aborigines made Australia*, Allen & Unwin, Sydney, 2011.

Gilfedder, Louise, James Kirkpatrick, Hans Wapstra and Annie Wapstra (eds), *The Nature of the Midlands*, Midlands Bushweb: Northern Midlands Council, Longford, Tas, 2003.

Griffiths, Tom, *Hunters and Collectors: The antiquarian imagination in Australi*a, Cambridge University Press, Melbourne, 1996.

Gunderson, Lance H, CS Holling and Stephen S Light (eds), *Barriers and Bridges to the Renewal of Ecosystems and Institutions*, Columbia University Press, New York, 1995.

Harrland, Karen, *Spinifex Baby*, Finch Publishing, Melbourne, 2014.

Holling, CS, *Adaptive Environmental Assessment and Management*, John Wiley & Sons, Chichester, UK, 1978.

Humann, Doug, 'A personal journey to innovation', in Figgis, Fitzsimons and Irvinge (eds), *Innovation for 21st Century Conservation*, pp 16–23.

Keon-Cohen, Bryan, *A Mabo Memoir: Islan kustom to native title*, Zemvic Press, Melbourne, 2013.

Kerezsy, Adam, *Desert Fishing Lessons: Adventures in Australia's rivers*, UWA Publishing, Perth, 2011.

Low, Tim, *Where Song Began: Australian birds and how they changed the world*, Penguin Books, Melbourne, 2014.

Malthus, Thomas Robert, *An Essay on the Principle of Population*, published anonymously, 1798.

Mant, Alistair, *Intelligent Leadership*, Allen & Unwin, Sydney, 1997.

Markus, Nicola, *On Our Watch: The race to save Australia's environment*, Melbourne University Press, 2009.

Martin, Sarah, *Davis McCaughey: A life*, UNSW Press, Sydney, 2012.

Meadows, Donella H, Dennis L Meadows, Jørgen Randers and William W Behrens, *Limits to Growth*, commissioned by Club of Rome, Universe Books, New York, 1972.

Menkhorst, Peter and Frank Knight, *A Field Guide to the Manmmals of Australia*, Oxford Univeristy Press, Melbourne, 2001.
Norman, James, *Bob Brown: Gentle revolutionary*, Allen & Unwin, Sydney, 2004.
Pascoe, Bruce, *Dark Emu Black Seeds: Agriculture or accident?*, Magabala Books, Broome, WA, 2014.
Pulsford, Ian, Graeme L Worboys and Gary Howling, 'Australian Alps to Atherton connectivity conservation corridor', in Graeme L Worboys, Wendy L Francis and Michael Lockwood (eds), *Connectivity Conservation Management: A global guide*, Earthscan, London and Washington, DC, 2010, pp 97–105.
Pulsford, Ian, Graeme L Worboys, Jane Gough and Tim Shepherd, 'The Australian Alps and the Great Escarpment of Eastern Australia conservation corridors', in David Harmon and Graeme L Worboys (eds), *Managing Mountain Protected Areas: Challenges and responses for the 21st century*, Andromeda Editrice, in assocn with IUCN World Parks Congress - 2003 Durban, Italy, 2004, pp 106–14.
Ratcliffe, Francis, *Flying Fox and Drifting Sand: The adventures of a biologist in Australia*, Chatto & Windus, London, 1938.
Richardson, Graham, *Whatever it Takes*, Bantam Books, Sydney, 1994.
Robin, Libby, *Defending the Little Desert: The rise of ecological consciousness in Australia*, Melbourne University Press, 1998.
Routley, Richard and Val, *The Fight for the Forests: The takeover of Australian forests for pines, wood chips and intensive forestry*, Research School of Social Sciences, ANU, Canberra, 1973.
Sattler, Paul S, and Rebecca D Williams (eds), *The Conservation Status of Queensland's Bioregional Ecosystems*, Environmental Protection Agency, Queensland government, Brisbane, 1999.
Schama, Simon, *Landscape and Memory*, Fontana Press, London, 1996.
Toffler, Alvin, *Future Shock*, Random House, New York, 1970.
—— *The Futurists*, Random House, New York, 1972.
Toyne, Phillip, *Growing up the Country*, Penguin Books, Melbourne, 1987.
Vaux, Kaye, 'The view east from Ongerup', in *The Bush Comes to the City*, Fitzgerald Biosphere Project, Jerramungup, WA, 1989.
Walsh, P, S Cowell, S Eccles and M Appleby, 'Beyond the Boundaries: Bush Heritage Australia's approach to multi-tenure conservation', in Fitzsimons, Pulsford and Westcott (eds), *Linking Australia's Landscapes*, pp 175–83.
Watson, Don, *The Bush: Travels in the heart of Australia*, Penguin Books, Melbourne, 2014.
Willis, James, *A Handbook to Plants in Victoria*, Melbourne University Press, 1970.
Wilson, EO (ed) *Biodiversity*, National Academy Press, Washington, DC, 1988.
Worboys, Graeme L, and Brendan Mackey, 'Connectivity conservation initiatives', in Fitzsimons, Pulsford and Westcott (eds), *Linking Australia's Landscapes*, pp 7–12.

Articles

Barnes, Simon, 'Without a sky full of insects there's no sky full of birds', Wild Notebook, *The Times*, 7 June 2014.
Beder, Sharon, 'Activism versus negotiation: Strategies for the environment movement', *Social Alternatives*, vol 10, no 4, December 1991, pp 53–56.
Brown, Bob, 'The quest for Drys Bluff: A view from Liffey', *Habitat Australia*, vol 18, no 4, August 1990.
—— 'Rhapsody of Oura Oura', *Ampersand Magazine*, no. 4, Spring 2011, pp 37–40.
'Bush buy back beats abuse', *Great Southern Herald*, 19 March 1997, p 13.
Button, James, 'Wild idyll', *Age*, 21 March 1997.
Dux, Monica, 'Growing up on troubled ground', *Age*, 13 December 2014.
Garnet, J Ros, 'The National Parks and the FNCV', *Victorian Naturalist*, vol 97, no 3, 1980, 130–34.
Goode, Dewar, 'Will Australia become another Sahara?', *Adelaide Mail*, 1935.
Laurie, Victoria, 'Inspire, Martin Copley', *Australian*, 2 August 2008.
Mackey, Brendan, Sandy Gilmore, Hugh Pringle, Paul Foreman, Linda Van Bommel, Sandra Berry and Murray Haseler, 'BioPrEP: A regional, process-based approach for assessment of land with high conservation value for Bush Heritage Australia', *Ecological Management and Restoration*, vol 11, no 1, April 2010, pp 51–60.
Naess, Arne, 'The shallow and the deep, long-range ecology movement. A summary', *Inquiry* (Oslo), vol 16, no 1–4, 1973, pp 95–100.
Pulsford, Ian, Graeme L Worboys, Jane Gough and Tim Shepherd, 'A potential new continental-scale conservation corridor for Australia combining the Australian Alps and the Great Escarpment of Eastern Australia conservation corridors', *Mountain Research and Development*, vol 23, 2003, pp 291–93.
Routley, Richard, 'Is there a need for a new, an environmental ethic?', *Proceedings of the XVth World Congress of Philosophy*, 1973, pps 205–10.
Traill, Barry, 'Populate or perish', *Australian Geographic*, vol 124, January–February 2015.
Wilson, Susan, 'Karen Alexander', *Wild Magazine*, July–August 2010, pp 18–19.

Online sources

Australian Government, Department of the Environment, 'History of the National Reserve System' <www.environment.gov.au/topics/land/nrs/science/scientific-framework>
Barlow, Nora (ed), *Autobiography of Charles Darwin 1809–1882*, Collins, London, 1958 <darwin-online.org.uk/content/frameset?viewtype=text&itemID=F1497&pageseq=1>
Bush Heritage Australia <www.bushheritage.org.au>
Bush Heritage Blogs <blog.bushheritage.org.au/blog/>
Bush Heritage Newsletters <www.bushheritage.org.au/news/headlines/news-2015>

Environment Protection and Biodiversity Conservation Act 1999 <www.comlaw.gov.au/Details/C2015C00422>

Hawke, RJL, *Our Country Our Future: Statement on the environment*, July 1989 <trove.nla.gov.au/work/16690946>

Hoben, ED, 'A national pleasure ground. The Kuring-gai Chase', *Sydney Morning Herald*, 14 April 1900 <trove.nla.gov.au/ndp/del/article/14305781>

Keating, PJ, 'Statement on the Environment', December 1992 <www.environment.gov.au/land/nrs/about-nrs/history>

Kosciuszko2Coast <k2c.org.au/K2C>

Mokany, Karel, D Friend, James Kirkpatrick and Louise Gilfedder, *Managing Tasmanian Native Pastures: A Technical Guide for Graziers*, Tasmanian Institute of Agricultural Research, University of Tasmania, 2006 <dpipwe.tas.gov.au/conservation/publications-forms-and-permits/publications/managing-tasmanian-native-pastures>

Rio Earth Summit, 1992 <www.un.org/geninfo/bp/envirp2.html>

Smith, Libby, 'Carnarvon Station: A history of European settlement since 1863', Bush Heritage Australia, 2003 <www.bushheritage.org.au/okaces-we-protect/queensland/carnarvon>

The Pew Charitable Trusts <www.pewtrusts.org/en>

Whitlam government environmental policy <whitlamdismissal.com/1972/11/13/whitlam-1972-election-policy-speech.html>

World Wide Fund for Nature (WWF) <www.wwf.org.au/about_us/wwf_australias_missions_and_goals/>

Woinarski, John, Barry Traill and Carol Booth, 'The Modern Outback: Nature, people and the future of remote Australia', Pew Charitable Trusts, 2014 <www.pewtrusts.org/~/media/assets/2014/10/themodernoutbackforweb.pdf>

Unpublished material

Cianchi, John, 'I talked to my tree and my tree talked back: Radical environmental activists and their relationships with nature', PhD thesis, University of Tasmania, 2013.

Hoare, John, 'The history of the National Trust', an outline of a talk given to Tickhill and District Local History Society, 20 November 2008.

Nicholson, Charles, 'Charles Darwin Reserve'.

Warren, Len, 'The woodlands of Charles Darwin Reserve: Overview 2012'.

Wommack, Kent, 'Confidential Report' on Bush Heritage, 2001.

Worboys, Graeme L, 'Conservation corridors and the NSW section of the Great Escarpment of Eastern Australia', paper presented at IUCN World Conservation Congress, Montreal, Quebec, Canada, 13–23 October 1996.

Private papers

Doug Humann
Margaret Robertson

Radio, film and television

'Bush Blitz', *Living Black*, NITV Television, 21 May 2013 <www.sbs.com.au/ondemand/video/30904387777/living-black-bush-blitz>

'Canberra Close Up: Phillip Toyne', interview with Alex Sloan, ABC Local Radio Canberra, 21 March 2012 <www.abc.net.au/local/stories/2012/03/21/3460338.htm>

'Feral cats rewrite the Australian story', Reporter: Gregg Borschmann, *Background Briefing*, ABC Radio RN, 12 October 2014 <www.abc.net.au/radionational/programs/backgroundbriefing/feral-cats-re-write-the-australian-story/5802204#transcript>

'Forestry peace deal: The end of another chapter', Andrew Fisher, *The Drum*, ABC Television, 20 October 2010.

'Generous donation saves Australian bushland', Reporter: Kerry Brewster, *The 7.30 Report*, ABC Television, 15 May 2008.

The Last Wild River, film by Paul Smith and Amanda Stark, Tas, (predates 1979) <trove.nla.gov.au/work/21844695>

The Nature of Things, presented by David Suzuki, CBC Television, Canada.

The Nature Series, presented by David Attenborough, BBC Television, various dates.

'Walmsley walks away from Earth Sanctuaries', Reporter: Prue Adams, *Landline*, ABC Television, 27 March 2005.

'Who are you? Martin Copley', ABC Local Radio Perth, 21 March 2013 <mpegmedia.abc.net.au/local/perth/201303/r1090292_13049116.mp3>

Interviews

(Conducted by Sarah Martin in 2014 and 2015)

Phillip Adams, David and Sue Akers, Karen Alexander, Shaun Ansell, Margaret Atkinson, Luke Bayley, Janice Bird, Greg Blake, Katrina Blake, Max Bourke, Keith Bradby, Allana Brown, Bob Brown, Mara Bún, Mike Chuk, Suzanne Connelly, Stuart Cowell, Peter Crosier, Troy Dare, Chris Darwin and Jacqui Courtney, Julian Davies, Tim Doherty, Kate Eddy, Sarah Eccles, Rod Fensham, Tim Fischer, Guy Fitzhardinge, Kate Fitzherbert, Paul Foreman, Chantal Fowler, Louise Gilfedder, Sandy Gilmore, Margaret Graham, Jack Green, Jodie Gunn, Ian Gust, Leanne Hales, Ian Hamilton, Julia Harris, Murray Haseler, Steve Heggie and Mel Sheppard, Judy Henderson, Richard Hines, Jeremy Hobbs, Doug Humann, Emma Ignjic, Jenny Kee, Adam Kerezsy, Seán Kerins, Penny King, Jamie Kirkpatrick, Denise Krake, Darren Larcombe, Sue McIntyre, Nathan Males, Nicola Markus, Jimmy Morrison, Steve Morton, Stephen Murphy, Charlie Nicholson, Glen Norris, Gerard O'Neill, Sally Port, Ian Pulsford, Peg Putt, Jim Radford, Margaret Robertson, Robert Rosen, Renee Rossini, Don Royal, Angela Sanders, Paul Sattler, Peter Saunders, Verna Simpson, Simon Smale, Paul Smith, Brian and Diana Snape, Hugo Spooner, Annette Stewart, Louise Sylvan, Victoria Templeton, Bill Thompson, Max Tischler, Phillip Toyne, Barry Traill, Keith Tuffley, Peter Unmack, Tom Vigilante, Phillippa Walsh, Len Warren, Dave Whitelaw, John Williamson, Chris Wilson, Roewen Wishart, Kent Wommack.

Glossary

ACF **Australian Conservation Foundation.** Established in the mid-1960s as a national conservation body with the commitment to achieve a healthy environment for all Australians.

Allens. Legal firm that began as Allen Allen and Hemsley, became Allens Arthur Robinson, and is now Allens Linklaters. Allens has provided pro bono legal advice to Bush Heritage for many years.

Anchors in the Landscape. Conservation strategy developed by Bush Heritage in 2005, which focused on building conservation communities within a broader landscape.

Australian Bush Heritage Fund. Original name for Bush Heritage that described the organisation's objectives. In 2007 the name was changed to 'Bush Heritage Australia'.

AWC **Australian Wildlife Conservancy.** Set up by Martin Copley in 1990 with a focus on protecting indigenous fauna.

Beyond the Boundaries. Concept developed by Bush Heritage in the early 2000s to express its commitment to working with all stakeholders in the particular regions in which Bush Heritage properties were situated.

Bush Heritage Australia. Generally known as 'Bush Heritage'.

CALM **Conservation and Land Management.** West Australian government agency.

CAR **'Comprehensive, Adequate and Representative'.** Framework used by the NRS program to determine land suitable for conservation purposes.

EPBCA ***Environment Protection and Biodiversity Conservation Act 1999.*** This Act established a legal framework for the protection of listed threatened ecosystems, plants and animals.

GA **Greening Australia.** Established in 1982 by the Labor government to deal with Landcare issues and to plant a billion trees. It has collaborated with Bush Heritage on a number of conservation projects, particularly in Western Australia.

Healthy Country Planning. Process based on the Open Standard for the Practice of Conservation that is designed to support Aboriginal Traditional Owners and custodians in planning for the protection of country, culture and community.

ILC **Indigenous Land Corporation**. Established in 1995 by the Australian government to assist Aboriginal people to acquire and manage land. It is the primary decision-making body for the 25 per cent of Australia for which Aboriginal people have title at law. It receives an annual payment of over $45 million from investment returns on the Aboriginal and Torres Strait Islander Land Account to run its programs.

IPA **Indigenous Protected Area.** An area of Aboriginal-owned land or sea where traditional owners and custodians have entered into an agreement with the Australian government to promote biodiversity and cultural resource conservation.

K2C **Kosciuszko to Coast.** Ambitious connectivity project in central New South Wales, similar in size and reach to the Gondwana Link project in Western Australia.

MOU **Memorandum of Understanding.** Document recording the intentions, responsibilities and understandings between Bush Heritage and a partner organisation that is the basis for a partnership of mutual respect and trust.

NGO **Non-government organisation.** Bush Heritage is one of many NGOs in the conservation sector.

NPWS **National Parks and Wildlife Service.** Originally set up to deal with conservation throughout Australia, but later each state developed its own agency to deal with local issues. The name still applies in South Australia, and was current in NSW until the early 2000s, when it became the Department of Environment and Conservation.

NRS **National Reserve System.** Program established by the Australian government in the early 1990s to support the private purchase of significant land for conservation purposes. It provided two dollars for every dollar raised by a participating organisation to encourage land to be purchased for conservation purposes.

TCT **Tasmanian Conservation Trust.** Established in 1968, it supported Bush Heritage in its early years.

TfN **Trust for Nature.** Established in Victoria under Liberal premier Dick Hamer, to support landowners wishing to protect native vegetation. Covenanting arrangements enable landowners to sell their properties, safe in the knowledge that the conservation values of the property will be maintained.

TLC **Tasmanian Land Conservancy.** Land-buying NGO in Tasmania which grew out of Bush Heritage's departure from Tasmania.

TNC **The Nature Conservancy.** An international conservation non-profit organisation, founded in 1951 in the US. The Australian branch was founded in 2002.

TWS **Tasmanian Wilderness Society.** Set up in 1976 amid concerns about the destruction of the Tasmanian forests; it led the Franklin River protests.

UTG **United Tasmanian Group.** The world's first Green political party, established in 1972.

VNPA **Victorian National Parks Association.** Established in 1952 to protect Victoria's unique natural heritage, and to educate and empower the Victorian community in supporting its goals. Doug Humann was the Executive Director before taking up the role as CEO at Bush Heritage.

WWF **World Wide Fund for Nature.** Began in 1947 as the Conservation Fund, was renamed World Wildlife Fund by Max Nicholson in 1961, and later became World Wide Fund for Nature, retaining its distinctive initials. It is the world's largest conservation organisation with over 5 million supporters worldwide.

Bush Heritage Australia executive and donors

Board members

Note: This listing is alphabetical.

Karen Alexander	1996–2001 (President, 1996–2000)
Greg Blake	1993–1995
Dr Bob Brown	1992–1997 (President, 1992–1996)
Mara Bún	1998–2007 (Vice-President, 1999–2000, Treasurer, 2002–2007)
Nick Burton Taylor	2013– (Treasurer, 2015–)
Janice Carpenter	2001–2006
Mike Chuk	2010–2013
Phillip Cornwall	2015–
Clem Doherty	1999–2000
Rick Farley	2001–2006
Penelope Figgis	1995–2003
Dr Guy Fitzhardinge	2002–2010
Louise Gilfedder	1998–2007 (Vice-President, 2000–2006)
Alastair Graham	1991–1993 (Treasurer)
Chris Grubb	2011–
Judy Henderson	1991–1993 (Secretary, 1992–1993), 1995–1998
Jeremy Hobbs	1992–1993
Dr Pierre Howitz	1998–2002 (filled temporary vacancy, 1998–1999)
Penelope King	1991–1992 (Secretary), 1993–1995
Prof Michelle Leishman	2015–
Leanne Liddle	2009–
Dr Sue McIntyre	2007–2015
Ray McKendrick	1993–2000 (Secretary, 1995–1997)
Dr Steve Morton	2003–2013 (Vice-President, 2006–2009)
Andrew Myer	2007–2016 (Vice-President, 2009–2016)
Dr Rebecca Nelson	2014–
Prof Hugh Possingham	2015–
Margaret (Peg) Putt	1991–1997 (Vice-President, 1991–1996)
Hutch Ranck	2006–2015
David Rickards	2006–2015 (Treasurer, 2007–2015)
Margaret Robertson	1994–2000
Robert Rosen	1993–2000 (Treasurer, 1994–2000)
Susan Rundle	1991–1992
Paul Sattler	2000–2002
Carol Schwartz	2004
Lincoln Siliakos	1996–1998 (Vice-President, 1996–1998)
Verna Simpson	1991–1992, 1993–1998
Libby Smith	1999–2006 (Vice-President, March 2000 – Nov 2000, Acting President Nov 2000 – March 2001)
Louise Sylvan	2009– (President, 2009–)
Phillip Toyne	2000–2009 (President, 2001–2009)
Keith Tuffley	2006–
Dr Brian Whelan	2000–2006
Alexis Wright	2006–2007

Company secretaries

Doug Humann	1997–2011
Chantal Fowler	2011–2013
Gerard O'Neill	2013–

Public officers

Doug Humann	1997–2011
Gerard O'Neill	2013–

Chief executives

Doug Humann	1997–2011
Gerard O'Neill	2011–

Major donors

1990

Bob Brown

1991–1995

Herbert Compton
Stan & Irene Johanson
Alastair Morrison
Mullum Trust
Orica Ltd
Ruth Pallett
John Williamson

1996–2000

Joan Anderson in memory of Roslyn Jones
Stan & Irene Johanson
Eve Kantor & Mark Wootton
George & Joan Masterman
John T Reid Charitable Trusts

2001–2005

Margaret Andresen
The Bunning Family
Commonwealth of Australia National Reserve System
Chris & Jacqui Darwin
Diversicon Environmental Foundation
Mary Dryden in memory of Friedrich Bauer
The Judith Eardley Save Wildlife Association
Judy Henderson
Stan & Irene Johanson
Eve Kantor & Mark Wootton
Lotterywest
Catriona and Simon Mordant AM
Michael Myer & Dellarose Rubi-Baevski
The Nature Conservancy
Ian Reid
John T Reid Charitable Trusts
South Australian Government
Thyne Reid Foundation
The R.E. Ross Trust
Dick Smith Foods Pty Ltd
Wesfarmers Ltd

2006–2010

Jo-Anne & Greg Beirne
Terry & Caroline Bellair
Besen Family Foundation
C Carlton
Craig & Cora Carter
Commonwealth of Australia National Reserve System
Consolidated Minerals Ltd
George Dalton & Annelie Holden
Philip & Carolyn Emms
Environmental Trust NSW
Flight Centre Foundation
Geoff Giles
Chris & Gina Grubb
The Hutchinson Foundation
JB Hi-Fi & Staff
Ian & Nan Landon-Smith
Macquarie Group Foundation
Maxwell Family Foundation
The Miller Foundation Ltd
Catriona and Simon Mordant AM
The Myer Foundation
Sidney Myer Fund 2009 Commemorative Grants Program
The Nature Conservancy
Nature Foundation SA
Various Trusts managed by Perpetual Philanthropic Services
John T Reid Charitable Trusts
David Rickards in memory of Helen Rickards
The R.E. Ross Trust
Royal Automobile Club of Western Australia
Christine Schwob in memory of Dr Peter Maling Barrer
South Australian Government
Louise Sylvan in memory of Richard Sylvan
The Thomas Foundation
Keith Tuffley
Vincent Fairfax Family Foundation
Wesfarmers Ltd
Graeme Wood Foundation

2011–2015

Andyinc Foundation
Terry & Caroline Bellair
Bob Brown & Paul Thomas
Phillip Cornwell & Cecilia Rice
Lesley Cunningham & Elizabeth Stephens
Gordon Douglas
Flight Centre Foundation
Geoff Giles
Gras Foundation
Chris & Gina Grubb
Ian & Dianne Gust
Jaramas Foundation
JB Hi-Fi & Staff
Macquarie Group Foundation
Maxwell Family Foundation
The Miller Foundation Ltd
Sidney Myer Fund
The Nature Conservancy
The Ian Potter Foundation
PW Foundation Pty Ltd
Robert & Sue Read
David Rickards in memory of Helen Rickards
Beth & Phill Schultz
Rosalie Schultz
Louise Sylvan in memory of Richard Sylvan
Michael Tichbon
Donald Tyler
In memory of Judy Wheeler, Lyle Wheeler and Harry Wheeler
Michael & Margaret Williams

Bequest supporters

1999

Estate of Dr Alex Griffiths
Estate of Dr Richard Sylvan

2002

Estate of Steve Roach
Estate of Marjory Augusta West

2003

Estate of Pixie Patricia Cope
Estate of Susan Jane Davey
Estate of Keith Alexander Grieve

2004

Estate of Ken Foley
Estate of Leilani Penny Musgrave

2005

Estate of Josephine Gay Bell
Estate of Howard Charles Richmond Fogarty
Estate of Karine Hayhurst
Estate of Dr Frank Peters
Estate of Val Williams

2006

Estate of Judith Margery Haworth Ambler
Estate of Stella Rose Cheah
Estate of Mary Dene Layton
Estate of Gwynneth Anna Oxnam
Estate of Nancy Vernon Rankine
Estate of Carolyn Meadley Stuart

2007

Estate of Henry Graham Baldwin
Estate of Lindsay Frederick Hunt
Estate of Sylvia Simpson-Lyttle
Estate of Stanley Charles Wilfred Stokes

2008

Estate of Christopher Daniel Blowes
Estate of Colin Glasford Campbell
Estate of Agnes Elizabeth Jackson
Estate of Claire Annette Maher
Estate of Valerie Mulholland Plumwood
Estate of Mary-Anne Prior
Estate of Madeleine Gladys Wright

2009

Estate of Bruce Leonard Dover
Estate of Ian Douglas Foote
Estate of Ernest Harold Jones
Estate of Maureen Veronica LeBlanc
Estate of Maria Hester Osborne
Estate of Doreen Lila Perry
Estate of Lois Mary Radford
Estate of John Walter Nesbitt Riddell
Estate of Ian Gordon Ross
Estate of Lindsay Shield Whitham

2010

Estate of Joseph Thomas Bedford
Estate of Christopher George Clark
Estate of Thomas Rosauer

2011

Estate of John & Jennifer Barnett
Estate of Lawrence Joseph Myer Barzel
Estate of John Hampton Hartwig Beale
Estate of Ivor Russell Bell
Estate of James Edward Alexander Bond
Estate of Ralph Barclay Braun
Estate of Gweneth Benson Cunningham
Estate of Rupert Gerald Featherstone
Estate of Marion Belfrage Field
Estate of John Colahan Griffin
Estate of Janet Peta Linnell
Estate of Philippa Mary O'Neill
Estate of Carolyn Joy Parsons
Estate of Anne Elizabeth Raymond
Estate of Mollie Vindin Rowell

2012

Estate of Laurie Violet Abell
Estate of Teresa Bocking
Estate of Ian Gordon Murray
Estate of Jane Alison Schwind
Estate of Anthony Glynne Smith
Estate of Jocelyn Marie Van Haren

2013

Estate of John Atkinson
Estate of Violet Cleverdon Austin
Estate of Virginia Tyler Leland Bonawit
Estate of Margaret Mary Cowper
Estate of Mary Griffith
Estate of Heather Fairlie Gulline
Estate of Donald Murray Hart
Estate of Joan M Hastwell
Estate of Alison Wynne Hearn
Estate of Judith Catherine Iltis
Estate of Thomas Kenneth Macdonald
Estate of Jennifer Mary Morrison
Estate of Clement Skevington
Estate of Barbara Helen Thompson
Estate of John Frederick Turner

2014

Estate of Sheila Mary Box
Estate of Saxon Marjorie Condon
Estate of Ruth Eileen Devenney
Estate of Judith Gaskin
Estate of Dora V Lee
Estate of Merlie Ivy Merchant
Estate of Yvonne Joan Viner
Estate of John Weightman
Estate of Brian Williams
Estate of Terence Cornthwaite Wollaston

2015

Estate of Mary Attenborough
Estate of Patricia Sarah Barringer
Estate of Donald Graham Batchelor
Estate of Paul Bayley
Estate of Barbara Joan Beeson
Estate of Margaret Sophie Billings
Estate of Virginia Russell Bird
Estate of Robert Frederick Brown
Estate of Helen Margaret Curtis
Estate of Garth Dixon & Rosalind Stafford
Estate of Audrey Ethel Twynam Horn
Estate of Violet Mabel Lynch
Estate of Devitt Kevin McGrath
Estate of Aina Matilda Ranke
Estate of Allan John Rogers
Estate of Anne Dallas Salvado
Estate of David Shepherdson
Estate of Costance Elaine Siddons
Estate of Peggy Smart

Acknowledgments

There are many people who kindly shared their stories about Bush Heritage with me, too many to mention, but I want particularly to acknowledge Bob Brown, Phillip Toyne, Doug Humann, Karen Alexander, Gerard O'Neill, and all the reserve managers who were so generous with their time, and replied to my questions over lengthy sessions.

I am especially grateful to Gerard O'Neill, who, having decided to entrust me with the writing of this history, let me come to my own conclusions. He and Kate Fitzherbert have been incredibly supportive, and have patiently read several drafts without complaining, offering help and advice when needed. And many thanks to Simon Muniandy and Loreen Chambers who, knowing nothing about Bush Heritage, agreed to read a draft to give unbiased opinions – their comments were very helpful.

The story has been greatly enhanced by the indefatigable Kate Fitzherbert, who has sourced the wonderful photos from the Bush Heritage archives and elsewhere, and worked them so successfully into the narrative. I am also indebted to the team at NewSouth, Jean Kingett, who spent painstaking hours ironing out irregularities in the text, and Emma Driver, who has worked with the designer Susanne Geppert to create a beautiful book.

And, of course, I cannot thank Peter enough for entering so wholeheartedly into this wonderful project that has taken us on several extended camping trips to remote places. It has been a great gig, and without him it would not have happened, and this book would not have been written.

Index

C

I

J

K

L

M

N

Y